Ecological Research Monographs

Series editor
Yoh Iwasa

More information about this series at http://www.springer.com/series/8852

Mamoru Watanabe

Sperm Competition in Butterflies

Springer

Mamoru Watanabe
Graduate School of Life and Environmental Sciences
University of Tsukuba
Tsukuba
Ibaraki
Japan

ISSN 2191-0707 ISSN 2191-0715 (electronic)
Ecological Research Monographs
ISBN 978-4-431-55943-6 ISBN 978-4-431-55945-0 (eBook)
DOI 10.1007/978-4-431-55945-0

Library of Congress Control Number: 2015957997

Springer Tokyo Heidelberg New York Dordrecht London

Front cover: A female of the Japanese parnassian, *Parnassius citrinarius*, with mating plug feeding on the flower of the skevish, *Erigeron philadelphicus*. *Back cover: Left:* A pair of the silver-studded blue, *Plebejus argus*, copulating on the grass. *Center:* Males of the alpine black swallowtail, *Papilio maackii*, forming aggregations on soil by the roadside and showing puddling behaviour after a sudden shower in the summer. *Right:* A male of the pale clouded yellow, *Colias erate*, courting on the wing toward a perching female, as the female shows the mate refusal posture with open wings and elevated abdomen. (All photos taken in Shirouma, Nagano, Japan, by Mamoru Watanabe)

Printed on acid-free paper

Springer Japan KK is part of Springer Science+Business Media (www.springer.com)

Preface

Among insects, butterflies are exceptionally well suited to treatment as a group, although most of them are not pests to human beings. Many years ago Prof. Paul R. Ehrlich of Stanford University sent many reprints including "the population biology of *Euphydryas editha*" to me, then a graduate student, and pointed out that butterflies are suitable model insects for evolutionary ecology. Being large, active by day, and strikingly coloured, butterflies are probably better known to nonspecialists than any other insects are except dragonflies, as stated by Philip S. Corbet (1999), the famous odonatologist. Butterflies have an important role to perform in promoting awareness of the increasingly urgent need for biodiversity of the terrestrial ecosystem, such as being a basic component in the food chain during immature stages and in pollination during the adult stage, as well as being important in studies of evolution and coevolution. However, many people, particularly in Japan, know little about the biology and ecology of butterflies. Instead, the biogeography, distribution, phylogeny, and life history of most Japanese butterfly species have been clarified by researchers, collectors, and lovers of nature. The complete list of host plants for each Japanese butterfly species and the improved techniques for rearing larvae in cages have also been reported.

Butterflies, as well as dragonflies, beetles, and cicadas, are safe for Japanese children who enjoy capturing flying adults and rearing larvae indoors. Furthermore, the life cycle of the cabbage butterfly has to be learned by pupils in elementary school, but the spiritual enlightenment that butterflies offer humans might almost certainly become but a fading memory after childhood. Therefore, like that of dragonflies, butterfly evolutionary ecology is not so highly regarded among insect sciences but, rather, as the subjects of children's interests. However, the sperm competition theory has begun to be developed partly on the basis of clarifying the mechanism of butterfly mating behaviour.

This book is concerned primarily with the behaviour and ecology of butterflies. I have tried to systematize information about integrated mating behaviour from first encounters with both sexes to the onset of egg-laying behaviour. Although the material presented in this book is limited to several taxa, mainly in Papilionidae and

Pieridae, the inclusion of the main topics within the chapters follows broadly the butterflies with similar mating systems, particularly female polyandry.

The number of publications on the behavioural ecology of butterflies has increased in recent years. A major objective of this book is to present an appraisal and review of past and current literature and to ensure that any significant observation reported here can be traced unequivocally to its source. While preparing the present book, I have sometimes viewed with concern the appearance of new literature and, I wondered whether sufficient time and space would come with the incoming tide of information. Then my determination to complete this book was sustained by my belief in the necessity of choosing literature selectively.

Tsukuba, Japan Mamoru Watanabe

Acknowledgements

I could not have written this book without the opportunity provided by a visiting professor in Sweden that allowed me to start research on the behavioural ecology of butterflies pertaining to sperm competition. I am very grateful for the hospitality of Stockholm University in 1996. My special thanks go to Christer Wiklund, Sören Nylin, Arja Kaitala, Nina Weddel, Per-Olof Wickman and Bengt Karlsson. My stay in Sweden would not have been enjoyable without the many services and opportunities provided by the Royal Swedish Academy of Sciences and by the Japan Society for the Promotion of Science.

Thanks are also due to the American researchers Mark J. Scriber, Paul P. Feeny, Ronald L. Rutowski, Felix A.H. Sperling, Mark Deering and Wyne Wheling, all of whom supported me with English conversation and for English usage in my papers. Michael T. Siva-Jothy of the U.K. also supported my work on the butterflies as well as dragonflies. The research for this book involved combing through a plethora of publications on various aspects of butterfly ecology and evolution, although a large number of figures and tables focussed on Papilionidae and Pieridae, most of which are reproduced from the publications of my coworkers, graduate and undergraduate students, in my laboratory. Special thanks must go to Kiyohito Kobayashi, Hiromi Komiya, Toshiko Maeda, Fumihiko Todo, Shinya Ando, Yasuyuki Nakanishi, Ayako Nagata, Takahiko Ito, Koji Sato, Takako Oh'ura, Kazuyuki Muraoka, Miyuki Nishimura, Masayo Hirota, Minetaka Bon'no, Tasuku Imoto, Ayako Hachisuka, Nao Dainoue, Taihei Kobayashi, Masaru Hasegawa, Mari Kamikubo, Shuko Irie, Nayuta Sasaki and Tatsuro Konagaya. Mizuho Suzuki read an earlier draft.

I also thank Aiko Hiraguchi, who originally suggested *Sperm Competition in Butterflies* to accompany *Ecological Research Monographs* for Springer. Keisuke Ueda introduced us and encouraged me to write this book. Without them, this book would definitely not have happened.

To all my friends who study ecology, every one of you has been a great help for my pursuing butterfly ecology. Special mention must go to Masae Shiomi, Kazuo Nozato, Yoshito Suzuki, Masami Takagi, Yoshitaka Tsubaki, Naota Osaki,

Nobuhiko Suzuki, Toshiyuki Sato, Minoru Ishii, Yoshiomi Kato and Hisashi Fujii. Atsushi Kokubo, Kazuki Miyashita, Fusao Nakasuji, Toshitaka Hidaka, Makoto Numata, Keiji Kiritani, Yoshiaki Ito and Shun'ichi Iwao have supervised and encouraged me with many comments and suggestions on my work.

Last, I am most grateful to my wife, Keiko, for her support.

Contents

Chapter 1
Introduction

Sperm competition has been conceptualised, in the narrow sense, as a process of sexual selection defined as the competition, between the sperm of more than two males who have mated with the female, for fertilisation of the eggs in the female internal reproductive organs (Parker 1970). In insects, once a female has copulated, the sperm transferred from the male is stored and maintained unchanged in female sperm storage organs that have specialised morphology, mainly tissue, to provide both nutrition and a constant ionic environment for sperm survival throughout her lifespan (e.g., Villavaso 1975). Therefore, insect sperm must retain its fertilisation capacity throughout the lifespan of the female, and the females are able to fertilise eggs using the stored sperm for oviposition at any time. Typically, when a female oviposits, each egg passes down the oviduct via a special pouch near the exit of the sperm storage organs. Females can control the position of the eggs in their internal genitalia and have the ability to release exactly the adequate number of spermatozoa needed to fertilise the eggs (Walker 1980). The efficiency of sperm utilisation at fertilisation has been reported as being high because no sperm wastage occurs and a single spermatozoon is generally used to fertilise a single egg (Thornhill and Alcock 1983).

Because the females of most insects are likely to be polyandrous, when a female accepts copulating with another male before the stored sperm are exhausted, temporal and spatial overlap of sperm from different males must occur within the female sperm storage organs, resulting in sperm competition. Assuming each spermatozoon of every mated male has similar ability for fertilisation, similar longevity and so on, sperm competition is expected to favour so-called opposing adaptations in males that prevent future males from mating with the female or preempt the sperm stored by the females from previous matings (Simmons 2001). Eberhard (1996) pointed out the male morphological, physiological and behavioural traits that reduce or avoid the risk of sperm competition, such as guarding mates from rival males, deposition of a mating plug in the female genitalia, transfer of antiaphrodisiac substances and removal of rival sperm.

M. Watanabe, *Sperm Competition in Butterflies*, Ecological Research Monographs,
DOI 10.1007/978-4-431-55945-0_1

In the evolutionary ecology of butterflies, reproductive parameters such as mating and oviposition behaviour with both paternal and maternal investment have been subjects of intensive investigation for their fitness (Roff 1992). Considerable variation in the conspicuous courtship behaviour across butterfly species has been demonstrated in relationship to phylogeny or on the basis of physiology. During the past 50 years, an impressive amount of information has been gathered regarding the ethology, behaviour and ecology of butterflies, and recently this information was made available in the form of a comprehensive review (Simmons and Siva-Jothy 1998). For example, UV reflectance on the wing surface of several pierid butterflies is a cue for identifying their mates (Obara 1970). Dorsal hindwing colouration is known to be a male characteristic used by females in their choice of a male (Rutowski and Rajyaguru 2013). Some pheromones are used for courtship behaviour in many species of butterflies, such as sulphur butterflies (Silberglied and Taylor 1978). Hair-pencils extruded from the tip of the abdomen of the males in some species of Danaidae are used to distribute volatile smells for courtship behaviour (Feltwell 1986).

When the female accepts the courting male, copulation follows, being inactive in every species. After the copulation is completed, the female has received a single spermatophore including sperm in the bursa copulatrix. Because every spermatophore shell remains in the bursa copulatrix during her lifespan, it is easy for researchers to count the number of spermatophores by dissection of the female abdomen. Each spermatophore is the result from a single successful copulation, indicating that the number of spermatophores equals the number of matings in the female. Dissection of wild-caught females to count the number of spermatophores in the bursa copulatrix has clarified that females of many species mate more than once (Burns 1968). However, for a long time it was thought that females mated only once, because species-specific mate refusal behaviour or a mating-plug phenomenon, the sphragis, which is put in the tip of the abdomen of mated females, is frequently observed in several species. Now, spermatophore counts are routinely used to assess female mating history in the wild.

The reproductive success of each sex is determined not only by the quantity of offspring that individuals of the subsequent generation produce, but also by the quality of those offspring. Although there is a trade-off between the size and fitness of offspring, female butterflies may maximise their reproductive success by laying as many eggs as possible, so long as these eggs are a certain minimum size. Research on butterfly eggs has addressed the effects of interspecific and intraspecific differences in adult size, as well as female age. Egg size or egg mass or both tend to decrease with increasing female age when the female has mated only once (Kimura and Tsubaki 1985). On the other hand, females do not seem to live long enough to lay all the eggs they produce (Watanabe and Nozato 1986). Information on the number of eggs laid throughout the lifespan can be used to assess the total investment of females in their offspring. Females utilise materials transferred by males during mating in egg production, in addition to the nutrients on which the larval stages feed.

A spermatophore is a protein-rich capsule that encloses the sperm and is digested by the female after copulation. Such substances not only affect vitellogenesis but also her somatic maintenance, with possible effects on longevity (Boggs and Gilbert 1979), and also increase fecundity (Watanabe 1988). In addition, increase in the genetic diversity of her offspring could be accomplished by multiple matings. Therefore, females tend to mate repeatedly to increase their offspring during their lifespan, resulting in the occurrence of sperm competition within their reproductive organs among their mates after copulation, as well as male–male competition behaviour before mating.

This book is intended as an introduction to sperm competition in butterflies related to the multiple matings of females. The lifetime mating frequency in females is rather stable among generations, years and local populations (Watanabe and Ando 1993), although the population density, operational sex ratios and the courtship activities of males are variable in time and space. In general, excess males fly about searching for females to mate, the so-called biased operational sex ratio, and it is likely that they persistently court females when they encounter them. For females, on the other hand, male courtship is sometimes harassment. Consequently, copulation behaviour imposes a cost for the females because of the time required for oviposition and feeding. There must be a trade-off between the cost of mating and the number of matings in females, showing a stable and optimal lifetime mating frequency.

Because females mate multiple times throughout the oviposition activity period, competition among males as to which male fertilises the eggs that she lays is common. Then, polyandrous females seem to force males to control the female's oviposition to ensure that they father the offspring. However, mate guarding behaviour, such as a territory around the oviposition substrates, is not effective: the male cannot accompany the female during her oviposition activity because the host plants are highly dispersed, and, as a consequence of their reproductive physiology and morphology, females have to perform spatial risk-spreading by extruding their eggs over different habitats. This risk-spreading may explain some behavioural and life history traits of insects, including butterflies, and should be a particularly useful strategy if a species breeds in highly unpredictable habitats (Koch and Suhling 2005). In addition, the outer shape of butterflies, their genital morphology, does not allow contact guarding in which the male is physically connected to the female during oviposition.

In addition to the genital morphology, sperm migration from the spermatophore to the spermatheca after copulation does not allow males to directly control the fertilisation of eggs in the female reproductive organs. Only sperm that has arrived in the spermatheca is used for fertilisation, primarily depending on the efficiency of sperm migration. Therefore, when the mated female mates again, sperm competition might arise in the spermatheca, beyond the influence of the male. However, after copulation the mated female has a considerable period in which she does not accept mating. Because a large spermatophore in the bursa copulatrix delays the repeat mating of the mated female (Sugawara 1979), one of the male's strategies to inhibit re-mating of his mate is an increase in spermatophore size, in which the female could receive much more nutrient.

In butterflies, two distinct types of sperm in the spermatophore have been known to transfer from male to female: one is eupyrene sperm and the other is apyrene sperm. The former is used for fertilisation, and the latter has no nuclear body. There seems to be an excess number of sperm of each type in a single spermatophore, when compared with the lifetime female fecundity and the number of eggs the female actually lays in her lifetime. In addition, a female could receive much more sperm, including apyrene sperm, because they mate repeatedly. Several hypotheses why apyrene sperm is produced in males, transferred to females and then migrates together with eupyrene sperm to the spermatheca, have been proposed on the role of apyrene sperm in sperm competition.

In Chap. 2, empirical observation on the mating behaviour of butterflies, from the encounter of the sexes to separation after copulation, is briefly described. Courtship display performed by males seemed to indicate female mate choice in sexual selection. The operational sex ratio shows an excess of sexually mature males in every local population, resulting in various mate refusal behaviours shown by mated females. Differences in the observation of continuous activity performed by a given individual in high flying behaviour did not confirm the repeat mating behaviour of the mated females and the number of matings. Therefore, female polyandry has been understood because the number of spermatophores in the bursa copulatrix could be counted.

In Chap. 3, the morphology of the reproductive organs in both sexes is described in relationship to copulation behaviour and sperm movement. The male ejaculate is structured as the spermatophore in the bursa copulatrix including eupyrene and apyrene sperm. Then, after the termination of copulation, sperm migrate from the bursa copulatrix to the spermatheca.

The ultimate purpose of the female is to increase the number of her offspring. Chapter 4 concerns egg development and oviposition, but does not describe deposition behaviour on the host plants in detail. After eclosion, ovarian dynamics in females is focussed in relationship to maternal and paternal investment, to the spermatophores and accessory substances transferred. Thus, the significance of multiple matings in females is discussed.

When females utilise male investment as well as sperm for oviposition, the ejaculation substances before and after copulation have to be considered. Chapter 5 describes the male investment with sperm. Paternity guards using the spermatophore are also mentioned.

In Chap. 6, possible sperm competition after multiple matings of the female is examined. Sperm transferred from a male will encounter severe competition when the male is more than the second male to mate with the female. On the other hand, when a male mates with a virgin female, his sperm will be affected by the sperm derived from the second male to mate. The evolution of strategic ejaculation by males is discussed.

Although the fitness of the female that mated multiply could be increased, females might prefer the sperm of males that increases the fitness of the females, for example, males that produce large spermatophores. Thus, sperm manipulation in females is discussed in Chap. 7. The phenomenon of cryptic female choice is also considered.

In Chap. 8, the role of apyrene sperm is discussed. It can be suggested that the apyrene sperm might encompass the final conflict for sperm competition among males.

References

Boggs CL, Gilbert LE (1979) Male contribution to egg production in butterflies: evidence for transfer of nutrients at mating. Science 206:83–84

Burns JM (1968) Mating frequency in natural populations of skippers and butterflies as determined by spermatophore counts. Proc Natl Acad Sci USA 61:852–859

Eberhard WG (1996) Female control: sexual selection by cryptic female choice. Princeton University Press, Princeton

Feltwell J (1986) The natural history of butterflies. Croom Helm, London

Kimura K, Tsubaki Y (1985) Egg weight variation associated with female age in *Pieris rapae crucivora* Boisduval (Lepidoptera: Pieridae). Appl Entomol Zool 20:500–501

Koch K, Suhling F (2005) Do behavioural and life-history traits vary with mate-guarding intensity in libellulid odonates? Can J Zool 83:1631–1637

Obara Y (1970) Studies on the mating behavior of the white cabbage butterfly, *Pieris rapae crucivora* Boisduval. III. Near-ultra-violet reflection as the signal of intraspecific communication. Z Vergl Physiol 69:99–116

Parker GA (1970) Sperm competition and its evolutionary consequences in insects. Biol Rev 45:525–567

Roff DA (1992) The evolution of life histories: theory and analysis. Chapman & Hall, New York

Rutowski RL, Rajyaguru P (2013) Male-specific iridescent coloration in the pipevine swallowtail (*Battus philenor*) is used in mate choice by females but not sexual discrimination by males. J Insect Behav 26:200–211

Silberglied RE, Taylor OR Jr (1978) Ultraviolet reflection and its behavioral role in the courtship of sulfur butterflies *Colias eurytheme* and *C. philodice* (Lepidoptera, Pieridae). Behav Ecol Sociobiol 3:203–243

Simmons LW (2001) Sperm competition and its evolutionary consequences in the insects. Princeton University Press, Oxford

Simmons LW, Siva-Jothy MT (1998) Sperm competition in insects: mechanisms and the potential for selection. In: Birkhead TR, Møller AP (eds) Sperm competition and sexual selection. Academic Press, San Diego, pp 341–434

Sugawara T (1979) Stretch reception in the bursa copulatrix of the butterfly, *Pieris rapae crucivora*, and its role in behaviour. J Comp Physiol 130:191–199

Thornhill R, Alcock R (1983) The evolution of insect mating systems. Harvard University Press, Cambridge

Villavaso EJ (1975) The rôle of the spermathecal gland of the boll weevil, *Anthonomus grandis*. J Insect Physiol 21:1457–1462

Walker WF (1980) Sperm utilization strategies in nonsocial insects. Am Nat 115:780–799

Watanabe M (1988) Multiple matings increase the fecundity of the yellow swallowtail butterfly, *Papilio xuthus* L., in summer generations. J Insect Behav 1:17–29

Watanabe M, Ando S (1993) Influence of mating frequency on lifetime fecundity in wild females of the small white *Pieris rapae* (Lepidoptera, Pieridae). Jpn J Entomol 61:691–696

Watanabe M, Nozato K (1986) Fecundity of the yellow swallowtail butterflies, *Papilio xuthus* and *P. machaon hippocrates*, in a wild environment. Zool Sci 3:509–516

Chapter 2
Historical Perspective

Keywords Accessory glands • Aedeagus • Territory • Age • Bursa copulatrix • Copula duration • Ductus ejaculatorius • Eupyrene sperm bundle • Hill-topping • Patrolling • Spermatophore

Male butterflies move over a large area in search of receptive females, infrequently returning to the same area, or patrol within a small area, frequently returning to the same locality. Factors affecting mate-locating behaviour have been mentioned traditionally in the abiotic environment, such as ambient temperature, as well as the biotic environment, such as body temperatures, lek assembly, and territorial perching. Visual cues can be used in the detection and discrimination of conspecific individuals by both sexes, because of their diurnal activities, although some species detect females by scent alone. Various courtship displays have been described in ethology. After adequate courtship behaviour by males, virgin females of most species will accept copulation. Until 30 years ago, it was believed that females in the majority of species mate only once, but that males have the capacity to mate more than once. A plug secreted by males of some species during copulation seemed to prevent further mating of females, supporting such female monogamy. Mate refusal postures of mated females when encountered by males also suggested maintaining female monogamy. Consequently, virgin females are receptive, showing little choice to mate, and mated females are faithful. However, counting the number of spermatophores in the bursa copulatrix of females has indicated that females of most species regularly mate more than once, because a single spermatophore is transferred from the male during a single copulation. Further matings occur after sperm reserves or nutrients are depleted, suggesting that females are not continuously available to mate. Although the sex ratio is unity, for most of the time there are more males than receptive females, and in such circumstances males are in competition to gain access to scarce receptive females. Therefore, female mate choice, particularly in mated females, would be favoured.

M. Watanabe, *Sperm Competition in Butterflies*, Ecological Research Monographs, DOI 10.1007/978-4-431-55945-0_2

Protandry

Mate location, courtship and copulation are the typical three stages of the mating process in butterflies. Both sexes have to find each other; then, the female must evaluate whether the male is suitable for her mate. If the female accepts, copulation and insemination might ensue, after which the female would start laying eggs.

Populations of temperate insects typically have discrete generations of reproductive adults once or a few times during the favourable seasons for the activities, depending on the number of generations they have in a year. The seasonal synchronisation of adult emergence causes most temperate insects to be protandrous, which means that males on average enter the reproductive population some time before the females. There are many examples in the literature showing that male insects are apt to contribute to the maintenance of protandry through rapid development during their larval stage (Forsberg and Wiklund 1988). Theories on protandry have been developed in terms of sexual selection acting either on males to maximise the expected number of matings (Iwasa et al. 1983) or on females to minimise the pre-reproductive period (Bulmer 1983). Therefore, male mating opportunity must be maximised by emerging before females, with late-emerging males being heavily penalised in evolution because of the small number of receptive females to be encountered. Although males emerging too early risk premature death before any receptive females have emerged, most males could have the chance of encountering a large number of receptive females when they adequately start to eclose. Therefore, the optimum time lag in the emergence time between sexes should be apparent.

Various mathematical models on protandry have been developed with the one key assumption that females mate only once (Fig. 2.1). Wiklund and Fagerstrom (1977) advanced the hypothesis that protandry is the optimal reproductive strategy of males, particularly in species maintaining female monandry. The protandry

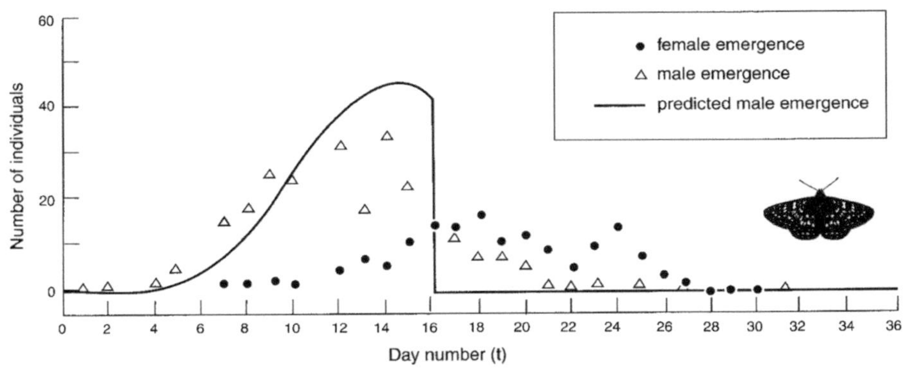

Fig. 2.1 Emergence curves of the checkerspot butterfly, *Euphydryas editha bayensis*, predicted for males using the model of Iwasa et al. (1983) and observed for both sexes. (From Boggs and Nieminen 2004)

Fig. 2.2 Pupae of the black-veined white, *Aporia crataegi*, on a twig of the host tree, *Prunus ssiori*

theory also focusses on species in which the generations are completely discrete (Singer 1982). As might be expected, because females are polyandrous in nature, protandry in respect to emergence is found in only a few species, in which males aggregate around female pupae that are close to emergence to mate with them. Watanabe (1978) observed that males of the black-veined white, *Aporia crataegi*, visited the pupae on the twig of the larval host tree, *Prunus ssiori*, to find females now eclosing (Fig. 2.2). Deinert et al. (1994) also reported that males of the Hewitson's long wing, *Heliconius hewitsoni*, search the larval host plants for female pupae, inspect these pupae regularly, then compete with each other for a position on a pupa and for access to the female as she ecloses. They could find female pupae not by adult appearance but by odour cues, as in the case of the common grass yellow, *Eurema hecabe* (Takanashi et al. 2001). For perching and guarding pupae with periodically assessing the developmental stage in *Heliconius* butterflies, Estrada and Gilbert (2010) pointed out that learning pupal location is a key factor in the mating strategy of the males.

Because selection for protandry is the strongest in species where females are monogamous (Fagerstrom and Wiklund 1982), it might be less important in species with an increased degree of female polyandry. Males of the monandrous peacock butterfly, *Inachis io*, set up and defend territories along forest edges several days before the arrival of females, after hibernation (Wiklund et al. 2003). On the other hand, multivoltine butterflies generally have overlapping generations, particularly towards late flying season. Considering that polyandrous females have no restraints placed upon them by males against subsequent matings, it can be assumed that there is no strong selection for protandry in these species. Zonneveld (1992) showed that with increasing degree of polyandry, protandry decreases, a result which is independent of any specific parameters. Wiklund and Forsberg (1991) pointed out that the absence of protandry in polyandrous species must be explained by sperm competition.

Encounter

Flight is essential to butterflies in looking for mates, food, and new habitats, to escape from predators, and, in females, for locating oviposition sites. These activities in turn affect the population structure and the operational sex ratio. The efficiency of flight and the capacity of butterflies to sustain flight activity is related to the initial thoracic temperature on takeoff and rates of heat production and loss. Watanabe and Imoto (2003) mentioned that the pale clouded yellow, *Colias erate*, inhabiting open grasslands, maintains a steady thoracic temperature during flight, despite the increasing heat from muscle contraction and radiation, and that air flow over the body (i.e., forced convection) effectively dissipates heat produced by muscle metabolism and radiation in the butterfly body. They show sustained and vigorous flight to be necessary for male mating success (Watanabe et al. 1997). It is thus important for survival in the pale clouded yellow that voluntary flight in open habitats results in the dissipation of excess heat.

To find a mate, male butterflies either fly in search of flying or alighted females or settle in particular localities in wait for flying females. For the former males, the flying behaviour might involve moving along the edge of hedges, woodland margins or forest rides, probably resulting in finding a mate. In some species, the males aggregate in special areas where some interaction within the species might be occurring. For the latter males, they usually occupy and defend the specific area such as forest gaps or sunflecks in the forest floor to increase the frequency with which they encounter receptive females, although there seem to be no obvious resources. Bergman and Wiklund (2009) mentioned three possible reasons why males in temperate forest-dwelling butterflies choose sunflecks or gaps as their perching sites, probably engaging territorial behaviour.

1. If such sunlit areas represent suitable rendezvous sites for males to encounter receptive females, those females must have a preference for visiting sunlit sites.
2. Because the sunlit areas allow males to keep a high body temperature, this may be beneficial for flight performance and the ability to take off rapidly and intercept passing receptive females on the wing.
3. Because sunlit areas are brighter than the surrounding forests, this might facilitate visual detection of receptive females when flying through the sunlit areas.

Within a territory, males of some species often perform conspicuous aerial interactions (Kemp and Wiklund 2001). Although the territoriality is influenced by weather, habitat type, vegetation structure, density of the local population and learning of the adults in the speckled wood butterfly, *Pararge aegeria* (Shreeve 1987), territorial perching males generally have higher mating probabilities than patrolling males (Vande Velde et al. 2011). Males exhibit display functioning as a preliminary fight against the intruders. The two males fly around each other followed by a chase without apparent physical attacks. Although bigger males are more successful in defending and overtaking territories (Rosenberg and Enquist 1991), the winner is usually the resident male. When a female flies into the territory, the resident male could detect her visually and chases her to copulate (Rutowski 1991). When the female flying into the territory alights nearby, the resident male also alights, walks to her and copulates (Wickman and Wiklund 1983).

Hill-topping behaviour is also an effective way of getting the sexes together (Vane-Wright 2003). Other features in the landscape could act as focal points, such as tall trees and gaps in a forest, both of which are occupied by males of some species that perch and wait for receptive females to come along, dashing out to investigate anything that moves. Hill-topping males defend perches on plants from which vantage points they can presumably detect the incoming females. While waiting for females, males perch for variable periods during the daytime on exposed branches and twigs of shrubs, as described by Alcock (2010). Therefore, various components of the hill-topping behaviour, such as perch site selection and site fidelity, as well as the contest behaviour, have to be considered separately in terms of their adaptive significance.

An alternative male tactic is patrolling, probably for many species, although the cost for searching females may be increased because of the higher proportion of time on the wing. Little site fidelity must be apparent. However, patrolling males could cover larger areas of the habitat and are more likely to encounter females. Then, males have to fly continuously through a large area, along tracks or forest margins, looking out for females. Watanabe (1979a) described the flyway of males in the swallowtail butterfly, *Papilio xuthus*, along the forest margins, which is the habitat for the host plants of the larvae.

Courtship Behaviour

Although sexual communication in butterflies involves the use of multiple signals at different phases of mate recognition and choice, constituting a complex scenario (Klein and de Araujo 2010), initial attraction usually takes place by visual channels (Scott 1972). The view is that butterfly wing pigment patterns are important for mate recognition and female choice. Females of the pipevine swallowtail butterfly, *Battus philenor*, preferentially mate with males that display a bright and highly chromatic blue on their dorsal hindwing (Rajyaguru et al. 2013). The information contained in the male visual signals must alter or control the female behaviour, and probably vice versa. If the development of the ornament is correlated with male genetic quality, females must gain the benefits in good genes for offspring, such as superior disease avoidance, as well as in receiving high nutrient-rich materials for themselves. However, Kemp et al. (2008) found that in the alfalfa butterfly, *Colias eurytheme*, ultraviolet (UV) coloration of males is not correlated with the quality of the spermatophore produced. In addition, pheromones are used for courtship behaviour to facilitate mating by males of some species, releasing from specialised scent organs such as hair-pencils, coremata and modified scales. Scent substances on the wings have been found in both sexes of the common grass yellow, *Eurema hecabe*, and both pheromones could be used in male discrimination (Takanashi et al. 2001), although only male sex pheromones elicit female mating behaviour in many butterfly species (Rutowski 1978).

Because of the differences in the relative positions of the female and male wings during the courtship flights in the pipevine swallowtail butterfly, *Battus philenor*, the perceived colour of the male's dorsal wing surface changes dramatically over a range of wavelength (Rutowski et al. 2010). For a given position of receiver and light source above a wing surface, both the brightness and the chroma of the reflection seen by the viewer must change as the wing moves during a wingbeat cycle. Silberglied and Taylor (1978) clarified that males of the alfalfa butterfly, *Colias eurytheme*, and of the clouded sulphur, *Colias philodice*, search for females visually, using UV reflectance, and that some pheromone is vital to the females for successful courtship.

In the pale clouded yellow, *Colias erate*, which showed the characteristics of patrolling species, Watanabe et al. (1997) showed that males can easily encounter females when they fly around the habitats, probably using visual cues. Irie and Watanabe (2009) also reported that newly emerged females are, in general, easily found by mate-searching males on the wing in the early morning. Such virgin females simply accept courting males and then copulate (Fig. 2.3). When the mated female flying accepts the courtship of the male, she alights on the vegetation, remaining motionless on the perch site. The male follows her, alights next to the female and bends his abdomen to the tip of the female's abdomen for copulation. When the female showed mate refusal behaviour because of recent copulation, males stopped their courtship behaviour and flew away from her (Hasegawa and Watanabe 2008). Consequently, in the daytime almost every flying female seems to be unavailable for searching males. Mated females often showed a pierid-specific mate refusal posture at the perching site (wings spread, abdomen nearly vertically upturned) or a so-called ascending flight followed by the courting males. However, dissection of females captured in the field shows that females mate a couple of times during their lifespan (Nakanishi et al. 1996), indicating that mated females do not always refuse males. Therefore, males have a chance to mate with experienced females as well as with virgin females.

There are species-specific ratios of encounter rates between the sexes. For example, virgin females of the small heath butterfly, *Coenonympha pamphilus*, search out male territories and show a lengthy solicitation flight, whereas mated females avoid being detected in male territories (Wickman 1986). The ringlet

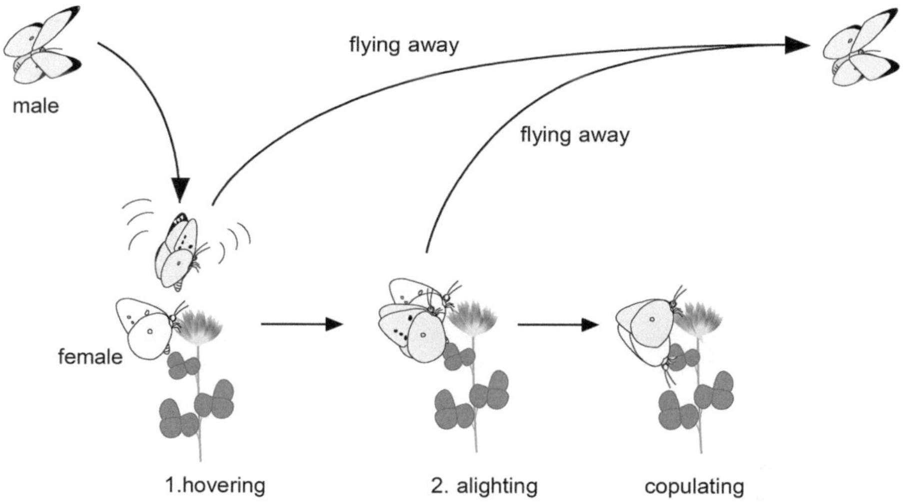

Fig. 2.3 Schematic representation of the courtship behaviour of the male pale clouded yellow, *Colias erate*, to the presented female (Hasegawa and Watanabe 2008). *1* A flying male visits the presented female and starts to hover in front of the female. *2* The male alights on the edge of the female's wings, walks down, and bends the abdomen towards the female; the male and the female then copulate. In each stage, when the males abandon copulation, they fly away from the female

butterfly, *Aphantopus hyperanthus*, also shows a similar shift in female behaviour after mating (Wiklund 1982). In the daytime, on the other hand, males of the pale clouded yellow, *Colias erate*, actively patrol on the wing, searching for females in open grasslands (Watanabe and Imoto 2003). They show powerful flying with rapid wingbeats, and cruise at a low height above the vegetation all day (except around noon, when the temperature is too high for flying) to search for females, with only a few intermittent periods of resting and feeding, while females perch in the vegetation and fly much less frequently. When males find a female perching or flying, they visit the female to start courtship behaviour. Males hover around the perching female, or near the flying female, inhibiting her flight course, probably using some pheromones.

In the Lepidoptera, there are many reports on the sequence of nuptial communication between males and females before copulation, including the female's response to the male's courtship behaviour (Wiklund 2003). As shown in Fig. 2.4, the female posture of exposing her abdomen from between the hindwings towards the male during the courtship behaviour has often been reported as a signal of female receptivity in butterflies such as the buckeye butterfly, *Precis coenia* (Scott 1975), the wood white, *Leptidea sinapis* (Wiklund 1977), the little sulphur *Eurema lisa* (Rutowski 1978), the alfalfa butterfly, *Colias eurytheme* and *Colias philodice* (Grula et al. 1980) and the pale clouded yellow *Coilas erate* (Watanabe and Irie 2011). On the other hand, in the pipevine swallowtail, *Battus philenor*, males that have greener hindwings might enjoy a greater mating success because the dorsal iridescent area of the hindwing in males is a sexual signal (Rutowski

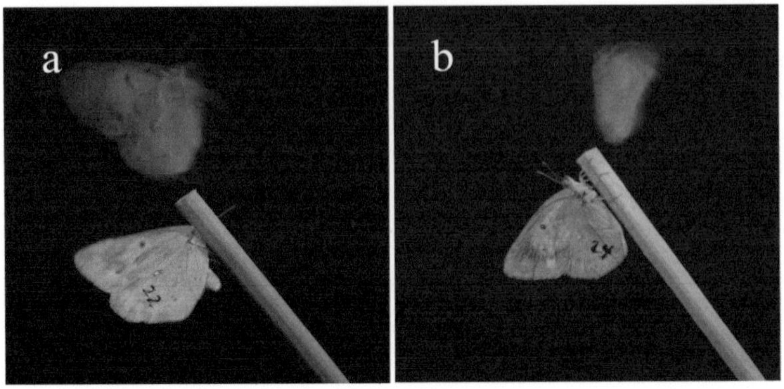

Fig. 2.4 Female response to the fluttering male in the pale clouded yellow, *Colias erate*. (**a**) The virgin female showed her exposed abdomen towards the male. (**b**) The unreceptive female did not expose the abdomen. (From Watanabe and Irie 2011)

et al. 2010). During aerial courtship manoeuvres in the pipevine swallowtail (Rutowski et al. 1989), the male positions himself below the female and flies up in front of her repeatedly, clearly displaying his dorsal wing surfaces to her. Rutowski and Rajayaguru (2013) clarified that ablation of male dorsal iridescence significantly reduced male mating success with virgin females.

Courtship proceeds along the behavioural sequence in which male and female recognise each other as a suitable mate. Successful copulation occurs only after the typical courtship behaviour is achieved (Silberglied and Taylor 1978). More specifically, male and female behaviour mutually release the next behaviour of both sexes. To evaluate female receptivity during courtship behaviour, males exploit the positive female behaviour. The female abdominal extension towards the male is an indication of female receptivity, and it may be a releaser of male alighting. Birch et al. (1989) demonstrated that when the females move off, away from male courtship behaviour, the males fly away from the female without any attempt for copulation. Therefore, butterfly females readily repel male attempts so that antagonistic coevolution between males and females does not escalate.

Copulation

Mating begins with the male positioning himself side by side with the stationary females. If the female, particularly of the pierid butterflies, is not receptive, she performs a specific mate refusal posture. If the females keeps her wings folded, leaving the tip of the abdomen physically reachable for the male, mating ensues (Watanabe and Irie 2011). Then, the male inserts his penis (aedeagus) into the vulva (vestibulum) of the female. To insert successfully, he has to engage her abdomen with a special locking device. At the tip of his abdomen there is a fixed (Fig. 2.5), hook-like structure (the uncus) with two lateral arms (valves) that close together to

Fig. 2.5 A pair of silver-studded blue, *Plebejus argus*, copulating on the grass

give a strong grip, as described by Vane-Wright (2003). Once firmly locked together, the tip of the aedeagus is then exerted between the valves to reach into the vestibulum, for entering the vagina proper. The male thus transfers his sperm to the female.

The mating ability of males on the day of eclosion is generally low, and the 1-day-old males start to search for females to copulate. Although the species-specific courtship behaviour performed by males is needed for females to accept the male in copulation, copulation by pairs of some swallowtail butterflies, mainly *Papilio* spp., *Byasa* spp. and *Battus* spp., can be artificially done by human hands, a so-called hand-pairing technique (West 1983; Scriber and Lederhouse 1988; Watanabe and Hirota 1999). The tips of the male and female abdomens are brought together, squeezed so the male opens his claspers and then the male is joined with the female (Fig. 2.6).

Even in the monandrous species, few virgin females are generally collected from a high density of the local population in the field, suggesting that females mate when encountering males soon after eclosion. In the black-veined white, *Aporia crataegi*, Watanabe (1978) observed that males are able to locate female pupae that are just emerging on the host tree of the larvae (Fig. 2.7). They scramble to approach the newly emerged females even though their wings are not yet fully enlarged and hardened. Some of the males even mount the pupa. The aim of males seems to be the first to copulate with the female, so that these females have no control over who will mate with them, although they can re-mate. Similar phenomena of such a 'pupal rape' were described in *Heliconius* spp. and the New Guinea yellow birdwing by Vane-Wright (2003).

Fig. 2.6 Schematic representation of the hand-pairing technique for the swallowtail butterfly, *Papilio xuthus*

Fig. 2.7 A male (*circle*) flying around a twig of the host tree of the black-veined white, *Aporia crataegi*. Pupae are on the defoliated twigs

There is actually very little variation in duration in copulation when the pair is not disturbed by conspecifics, or predators including birds, other insects and humans. Rutowski and Gilchrist (1986) suggested that duration of copulation might be relatively long as a result of the mechanical problems of filling the bursa copulatrix. Successful copulations in the swallowtail butterfly, *Papilio xuthus*, lasted about 1 h (Fig. 2.8), by which time the spermatophore had been structured in the bursa copulatrix and substantially filled with white creamy secretion and sperm (Watanabe et al. 2000). The copula duration in many species is at least an hour for

Fig. 2.8 Copulation of the swallowtail butterfly, *Papilio xuthus*

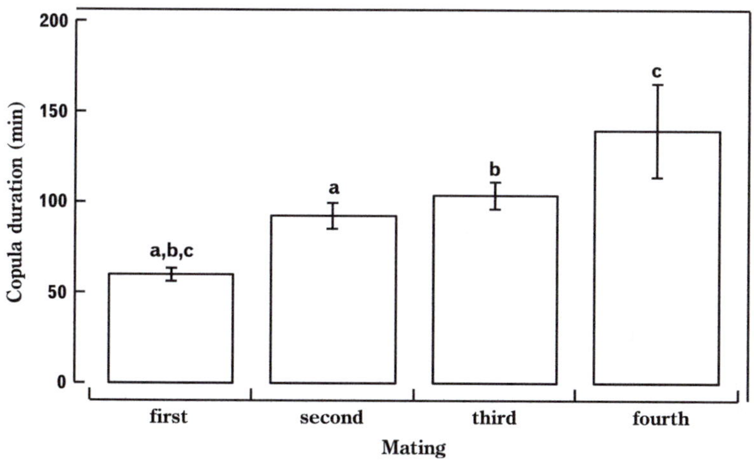

Fig. 2.9 Mean copula duration with males of the cabbage white, *Pieris rapae*, from the first to the fourth mating. *Bars* are standard errors. The same letters are significantly different at the 1 % level by Mann–Whitney *U* test. The number of matings exerts a positive effect upon copula duration by Kendall test (Watanabe et al. 1998b)

swallowtail butterflies (Watanabe and Hirota 1999) and for the cabbage white, *Pieris rapae* (Watanabe et al. 1998a), and sometimes much longer, as in the pale clouded yellow, *Colias erate*, because of harassment by the single males (Watanabe et al. 1997). Watanabe et al. (1998b) clarified that male cabbage whites increased copula duration with the number of matings (Fig. 2.9). Thus, the males that mated three times (they must be old at that time) copulate for more than 2 h. Caballero-Mendieta and Cordero (2013) also reported that the copulation duration for relatively large males of the monandrous pierid butterfly, *Leptophobia aripa*, was 83 min, 93 min and 491 min, for his first, second and third copulation, respectively. Kaitala and Wiklund (1995) showed that copulations with recently mated males of the green-veined white, *Pieris napi*, can last up to 24 h, suggesting that the increase in the number of matings may carry substantial time costs. On the other hand, the copula duration of virgin males of the monandrous small copper, *Lycaena phlaeas*, is about 15 min (Watanabe and Nishimura 2001), and for the monandrous speckled wood butterfly, *Pararge aegeria*, it is about 20 min (Vande Velde et al. 2011).

Substances are passed to the female continuously throughout copulation. This deposition represents approximately 7 % of the male's body mass for the cabbage white, *Pieris rapae* (Watanabe and Sato 1993) and the clouded sulphur, *Colias philodice*, or the alfalfa butterfly, *Colias eurytheme* (Marshall 1985). Spermatophore transmission from male to female during copulation must result in weight loss from the male body and the corresponding weight increase of the female body. Figure 2.10 shows that during copulation the weight increase in the female roughly

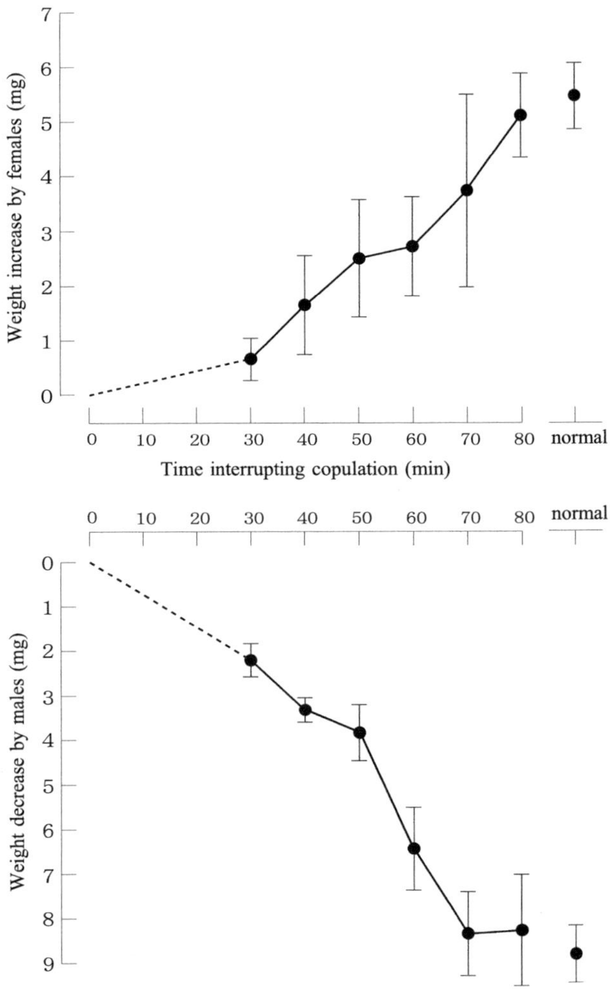

Fig. 2.10 Weight increase by mated females and loss by the mated males during interrupted and uninterrupted (= normal) copulations (±SE) in the cabbage white, *Pieris rapae* (Watanabe and Sato 1993)

coincides with the weight loss in the male. In an experiment with the cabbage white, there was no detectable weight loss by mated males 20 min after the onset of copulation, but thereafter male weight began to decrease. The loss was about 8–9 mg in 70 min or 80 min of interrupted copulation, each of which was not significantly different from that of normal copulations. The weight of the mated females also had not increased at 20 min after the onset of copulation, suggesting that little transfer from male to female occurred during the first 20 min of the copulation. The weight increase at 80 min of interrupted copulation was not significantly different from that of normal copulation. The mean wet mass of the spermatophore was about 5.8 mg, and the mean body mass of the virgin males and virgin females was 79.0 mg and 81.6 mg, respectively. Consequently, the mass of a spermatophore, plus the appendix bursa contents, represents 7.3 % and 7.1 % of the body mass of males and females, respectively. On the other hand, 0.83 % and 0.94 % of male body weight is invested in the ejaculate produced by the virgin male of the monandrous mountain white, *Leptophobia aripa* (Caballero-Mendieta and Cordero 2013) and the speckled wood, *Pararge aegeria* (Bissoondath and Wiklund 1995), respectively.

Although the duration of copulation in butterflies seems to be under male control (Wickman 1985), one of the possible costs of mating incurred by males in transferring spermatophores is time. Many butterflies try to decrease copula duration against the risk of predation (Rutowski 1984), although prolonged mating has a function other than spermatophore transfer alone irrespective of size. In pierid butterflies, differential transfer rates result from behavioural competition among males (Svärd and Wiklund 1988).

Thibout (1979) mentioned that copulation is not limited to a mere transfer of sperm from males to females, but is actually a complex process in which various behavioural sequences should be distinguished and followed by various physical effects on the female. Then, the important aspects of the copulation are as follows:

(a) Precopulatory sexual behaviour
(b) Introduction of the male genitalia into the female
(c) Transfer of sperm and the various male secretions into the bursa copulatrix
(d) Presence of the spermatophore and swelling of the bursa copulatrix
(e) Presence of sperm and the male secretions in the bursa copulatrix
(f) Activity of sperm in the spermatheca after sperm migration from bursa copulatrix to spermatheca

Spermatophore

After the onset of copulation, particularly for the swallowtail butterflies, *Papilio* spp., and pierid species, the bursa copulatrix, initially a thin, flat ellipse, grows rapidly in volume and weight, and becomes oval at about the time the spermatophore is structured (Fig. 2.11). For the cabbage white, *Pieris rapae*, the weight of the bursa copulatrix in virgin females is 0.84 mg. When the weight of the ejaculate

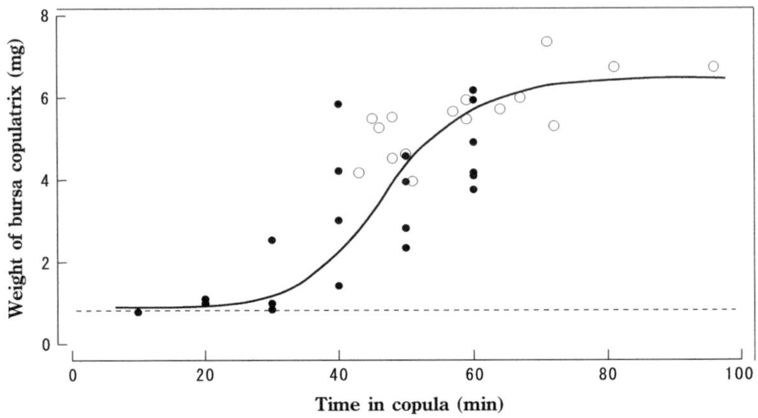

Fig. 2.11 Change in weight of bursa copulatrix of females after interrupted (*closed circles*) and uninterrupted (*open circles*) copulations in the cabbage white, *Pieris rapae*. *Dotted line* shows mean weight of the bursa copulatrix of virgin females. (From Watanabe et al. 1998a)

can be assessed by subtracting the weight of the bursa copulatrix of virgin females from the whole weight of the bursa copulatrix of mated females, there is no detectable change in the weight of ejaculate at 20 min after the beginning of copulation (Watanabe et al. 1998a). The weight of the ejaculate begins to increase thereafter. Then, during copulation, the male secretions from the ductus ejaculatorius and accessory glands are transferred serially to the bursa copulatrix. The ejaculate is white gel mainly observed on the tip of the penis 10 min after the beginning of copulation. There is no spermatophore capsule in the bursa copulatrix after 20 min in copulation. Then, at 45 min of copulation, the ejaculate weight is not significantly different from that observed in cases of uninterrupted copulation.

For the pipevine swallowtail, *Battus philenor*, the simplex of virgin males is a long tube and contains white, granular material and clear, pasty material (Sasaki et al. 2015). After the first mating, then, the simplex is shorter and lower in mass, and neither the white, granular substance near the aedeagus nor the clear pasty substance in other parts of the simplex are found in the simplex. Their relative position in the simplex requires that the white material must be placed first in the bursa copulatrix, followed by the clear substance. In the swallowtail butterfly, *Papilio xuthus*, spermatophore and accessory substances are transferred to the bursa copulatrix separately, and the accessory substance is located behind the spermatophore (Watanabe et al. 2000).

The spermatophore proper represents a balloon-like or a teardrop-shaped structure with a swollen corpus and a tube-like column with an opening at its tip (Fig. 2.12). In addition, gel-like accessory substances are transferred to the bursa copulatrix. Ejaculates including both spermatophore and accessory substances are physiologically costly for males to produce, with ejaculate mass ranging from 1.4 % in the speckled wood, *Pararge aegeria* (Svärd and Wiklund 1989) to 15.5 % in the Labrador sulphur, *Colias nastes* (Rutowski et al. 1983). Males of the green-veined

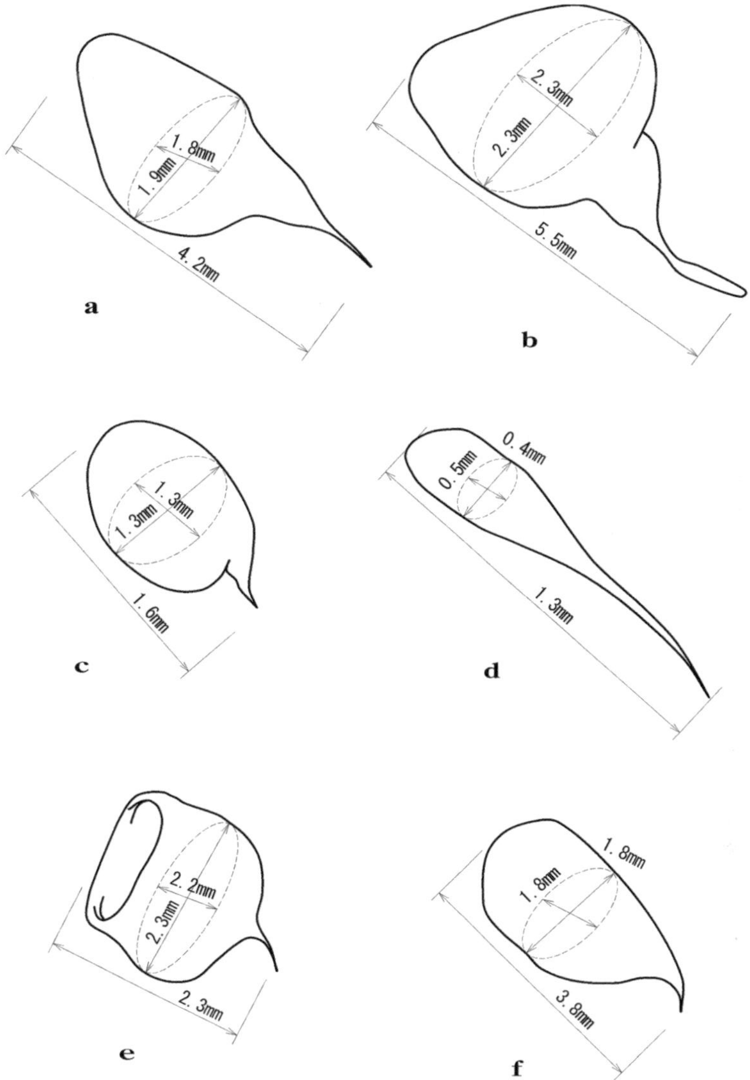

Fig. 2.12 An intact spermatophore just after termination of copulation in the common straight swift, *Parnara guttata* (**a**); the common yellow swallowtail, *Papilio machaon* (**b**); the cabbage white, *Pieris rapae* (**c**); the small copper, *Lycaena phlaeas* (**d**); the painted lady, *Vanessa cardui* (**e**); and the Japanese nymphalid butterfly, *Lethe diana* (**f**)

white, *Pieris napi*, have been observed to transfer a spermatophore corresponding to 23 % of their body mass, and its size is correlated with male body size (Fosberg and Wiklund 1989). The fresh weight of the spermatophore just after termination of copulation is about 6 mg in the swallowtail butterfly, *Papilio xuthus* (Watanabe et al. 2000) and the cabbage white, *Pieris rapae* (Watanabe et al. 1998b).

Although in most invertebrates females are usually larger than males, in some butterfly species sexual size dimorphism increases with the degree of polyandry, with males being larger than the females (Wiklund and Forsberg 1991). Across the species, such a relative male size increases with female polyandry, in relationship to the positive association between spermatophore size and male size (Bissoondath and Wiklund 1997). Because the large spermatophore is beneficial by both delaying female re-mating and increasing female reproductive output, large-size males must have a high advantage over small-size males.

Because sperm is stored in the duplexes that unite to form the simplex, sperm must be transferred to the female after transfer of the clear pasty substance. Thus, sperm are the last materials transferred to the female reproductive tract in the cabbage white, *Pieris rapae* (Watanabe and Sato 1993) and in the alfalfa butterfly, *Colias eurytheme* (Rutowski and Gilchrist 1986). At this time, the spermatophore contains immobile eupyrene sperm bundles and an actively free apyrene spermatozoa in the swallowtail butterfly, *Papilio xuthus* (Watanabe and Sasaki 2010) and in the cabbage white (Watanabe et al. 1998a, b).

Sperm stay in the spermatophore for several hours after copulation, during which time eupyrene sperm bundles begin to disappear with an increase in free eupyrene spermatozoa in the swallowtail butterfly, *Papilio xuthus* (Watanabe et al. 2000). Sperm of the large white, *Pieris brassicae*, starts to migrate to the spermatheca 5.5 to 8 h after copulation (Tschudi-Rein and Benz 1990). The spermatophore must be a site of sperm maturation during the period (Osanai et al. 1987). Accessory gland products in the spermatophore from the male have been shown to function in sperm activation (Leopold 1976). Sugar content in the spermatophore might contribute to sperm survival during this period (Watanabe and Sato 1993).

In some species, a spermatophore contains nutrients for the female that may support somatic maintenance (Boggs and Gilbert 1979) or increase her fecundity (Watanabe 1988). Males that can produce larger spermatophores are expected to have greater fertilisation success, especially in the polyandrous species (Parker 1982), because large spermatophores increase the length of the refractory period of the female, that is, the time during which the female is unreceptive to male courtship following copulation (Sugawara 1979). In addition, males might gain sperm priority by transferring a large spermatophore in some species (LaMunyon and Eisner 1994). However, the advantage of high sperm number might be relatively small in butterflies because of evidence that in many species, including the swallowtail butterfly, *Papilio xuthus* (Watanabe and Sasaki 2010), sperm mixing in the sperm storage organ rarely occurs in females mated more than once, resulting in the bimodal distribution of paternity or in the precedence of the last male sperm.

Estimation of Age

Realised adult longevity is a crucial life history trait because it will influence how many times an individual can engage in reproductive activities. Longevity must be essential to increase fitness for both sexes. Thus, males as well as females can have the chance to mate repeatedly during a long lifespan, as the mortality rate increases by external factors such as predation, starvation or drought with age. Age-specific fecundity pattern in relationship to female longevity has been reported in the laboratory by Labine (1968). Because adults of species that have one generation in a year show ageing relatively similar to each other because of the similar emergence period, each adult collected in the field must not be so considered in age for study. However, in the case of species that have more than two generations in a year, adults from different generations could coexist from summer to early autumn, because of adult longevity longer than the length of the larval period, weather conditions and so on. Consequently, there are many adults of various ages flying during the active flying seasons.

Wiklund et al. (2003) stated that all other things being equal, selection for extended male longevity should be stronger in polyandrous mating systems in which males are allowed to mate with females throughout the female flight period than in monandrous mating systems in which male mating opportunity is considerably more restricted in time. Although females of some polyandrous species must become unreceptive for a number of days after mating, females receptive for copulation must appear throughout the flying season. In other words, because females mate multiply throughout their lives, males could increase the number of matings with such receptive females throughout their own lives. On the other hand, for monandrous species, male mating success may not be expected to be associated with lifespan because the time period of the appearance of receptive females is rather short. Therefore, the male mortality rate seems to be greater than the female mortality rate in monandrous species.

Applying a mark-and-release survey to field populations has clarified many population parameters including estimated longevity, by using the Jolly–Seber method and the Manly and Parr method for various butterfly species such as the checkerspot butterfly, *Euphydryas editha* (Ehrlich 1965) and the black-veined white, *Aporia crataegi* (Watanabe 1978), both of which have one generation in a year. Population parameters of the black swallowtail butterfly, *Papilio protenor* (Fig. 2.13) and of *Papilio helenus* in summer generations were surveyed with the consideration

Fig. 2.13 A male of the spangle, *Papilio protenor* (black swallowtail butterfly), feeding on the flowers of the bushkiller, *Cayratia japonica*. (Photograph by T. Kasui)

of age of each adult captured by Kiritani et al. (1984). Nozato et al. (1985) proposed a survivorship curve of adults of black swallowtail butterflies wherein low and high mortality is found during young and old ages, respectively. With the exception of species that pass the winter as adults, most butterflies commonly live 2 or 3 weeks, unless overcome by accident or killed by predators. Ehrlich and Gilbert (1973) reported an exceptional species; the longevity of the ethilia longwing, *Heliconius ethilla*, was 6 months in the tropics.

After eclosion, every butterfly begins to show age. The colours of body and wings start to fade. Watanabe and Ando (1993) suggested that UV reflectance for the cabbage white, *Pieris rapae*, decreases with age after emergence. In general, the wings become rubbed and frayed from encounters with other insects including butterflies, the complex structure of vegetation, spider webs, and all manner of little accidents. There is no opportunity for repair, because there is little or no cell division, and no growth in the adult butterfly, other than those processes involved with reproduction. Therefore, the butterfly will perish at least after some weeks. As the males of most species age, probably with the experience of matings, in general, they might produce smaller and smaller spermatophores. Then, one of several reasons for receptive females to reject a male is that he is old. Consequently, the reward for mating with an old male may simply be too low. Females need to be sure of the male's relative age, as well as recently mated males, mainly by noting the wing wear of males.

To clarify the internal condition of butterflies as well as population structure, the age of each butterfly collected in the field has to be estimated. Because colouration degrades with age and wing wear, on the basis of the degree of wing wear, several age groups can be classified for the swallowtail butterflies, *Papilio polytes* (Watanabe 1979b), *Papilio xuthus* and *Papilio machaon* (Watanabe and Nozato 1986), and the pierid butterflies, *Pieris canidia* (Ando and Watanabe 1992), *Pieris melete* (Ando and Watanabe 1993) and *Pieris rapae* (Watanabe and Ando 1993). The five age groups (FF, F, B, BB and BBB) can generally be classified. An adult of FF group is defined as a recently emerged one with intact wings and lustrous scales. In addition, the bright coloration in FF group might indicate a low probability of previous mating. Adults of F group have still fresh wings with fine tears but less lustrous scales. Adults of B group have wing tears with frayed scales. Adults of BB group show notched tears and highly frayed scales. Wings of adults of the BBB group are broken or have extensive tears and highly frayed scales (Fig. 2.14).

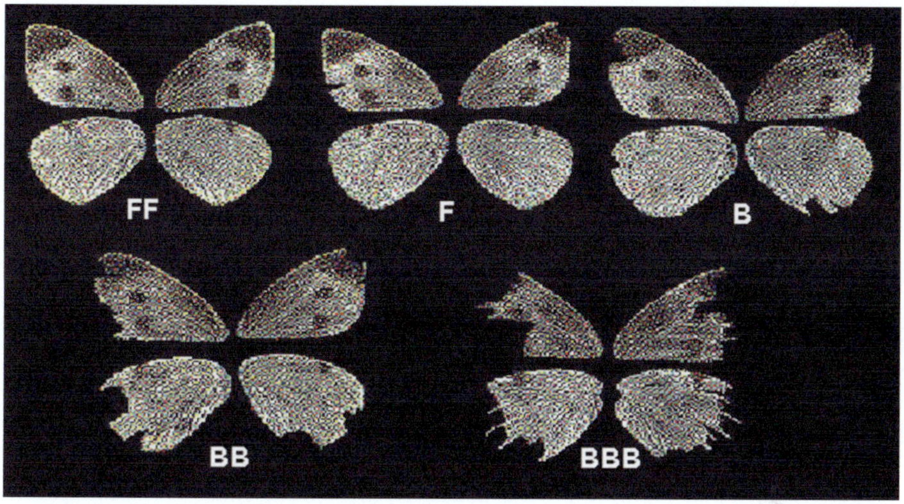

Fig. 2.14 Five age classes for the cabbage white, *Pieris rapae*

Lifetime Mating Frequency

Because a key determinant of the intensity of sexual selection is the variance in the reproductive success of individuals in natural populations, it is important to determine the lifetime mating number of both sexes. It had been believed that females of some species with a very short longevity usually mate only once, whereas females with a long period for oviposition will mate several times. The reason might be to replenish their store of sperm, to obtain more viable sperm, or to gain additional nutrients that will help to produce more eggs or allow somatic maintenance. Indeed, Vane-Wright (2003) stated that females of some milkweed butterflies mate frequently to obtain continual supplies of alkaloids, both for the protection of their eggs and for their own survival, because the alkaloids are highly unpalatable to many predators, including birds. Therefore, it is important to examine multiple matings in females that have mated multiple times to different males throughout their lifespan, that is, true polyandrous matings. Repeated matings, defined as mating the same male several times, are not common in butterfly species.

Mating success of males can only be documented by seeing them in copula. Consequently, male mating success in natural populations can often be described only by observing copulations, such as in lekking animals (Fiske et al. 1998) and damselflies (Fincke 1982). Males of most butterfly species probably mate as many times as they can. Some males copulate several times before they die, but unsuccessful males may die as virgins. Therefore, males would show a higher variance in reproductive success than the females. In addition, the expected number of matings by males correlates with the degree of polyandry exhibited by the species (Karlsson 1996); this contrasts with females, most of which mate at least once.

The lifetime mating success of wild females is often easier to assess than that of males. Females of some insects are wounded by male external spines during copulation, and these wounds can indicate mating history (Crudgington and Siva-Jothy 2000). In butterflies, however, females receive a spermatophore during a single copulation without any apparent injury by the male, and the remains of individual spermatophores can be found in the bursa copulatrix and counted even after they are drained of their contents (Burns 1968). The bursa copulatrix of females is balloon shaped, and it can expand to several times its original size depending on the number of spermatophores. Thus, female butterflies have considerable capacity for the storage of multiple spermatophores (Parker 1970).

Spermatophore counts of wild-caught female butterflies, which are routinely used to assess female mating history, indicate that the females of many species mate more than once (Lederhouse 1981). The reason for female polyandry has been believed to be that the sperm from their previous partners have been fully utilised. However, a number of studies have shown that they often re-mate before sperm exhaustion and that multiple mating occurs in the wild in many butterfly species (Stern and Smith 1960). Multiple matings, accordingly, seem to be the rule in most butterfly species, although females of the wood white, *Leptidea sinapis* (Wiklund 1977), the small copper, *Lycaena phlaeas* (Watanabe and Nishimura 2001) and the pipevine swallowtail, *Battus philenor* (Konagaya et al. 2015) are monogamous. However, the number of spermatophores in many studies was "the average mating number" for all females collected, ignoring female age.

The lifetime mating number of females should be estimated by the number of spermatophores in the bursa copulatrix, considering from that for the youngest to the oldest females. The mating number of females increases with the age group. Thus, the relationship between estimated female age and the number of spermatophores in the bursa copulatrix could estimate the lifetime mating number. The lifetime mating number of the black swallowtail butterflies (*Papilio helenus* and *Papilio protenor*) in a summer generation is 3 to 4 (Watanabe et al. 1986). For the swallowtail butterfly, *Papilio xuthus*, the lifetime mating number was about 3 for spring and summer generations (Watanabe and Nozato 1986). For the cabbage white, *Pieris rapae*, in the summer generation the lifetime mating number was about 3.5 for every year. The average number of matings in females of the polyandrous comma butterfly, *Polygonia c-album*, which is univoltine and overwinters in the adult stage, is 2.5 (Svärd and Wiklund 1989). Konagaya and Watanabe (2015) showed that the autumn-morph females of the common grass yellow, *Eurema mandarina*, mated once for their own somatic maintenance before overwintering and then re-mate for fertilisation in oviposition after overwintering. On the other hand, Tatar (1991) reported that the females of the pipevine swallowtail, *Battus philenor*, in early-spring and late-spring generations in California, contained 1.1 and 1.3 spermatophores, respectively. In summer generations of the pipevine swallowtail in Arizona, Konagaya et al. (2015) found that the first mating of females occurred very soon after eclosion but that the lifetime mating number was 1.4 without any change in the volume of spermatophores.

As shown in Fig. 2.15, the number of matings in the cabbage white, *Pieris rapae*, increased with age in each year, and the pattern of change is similar in all 3 years

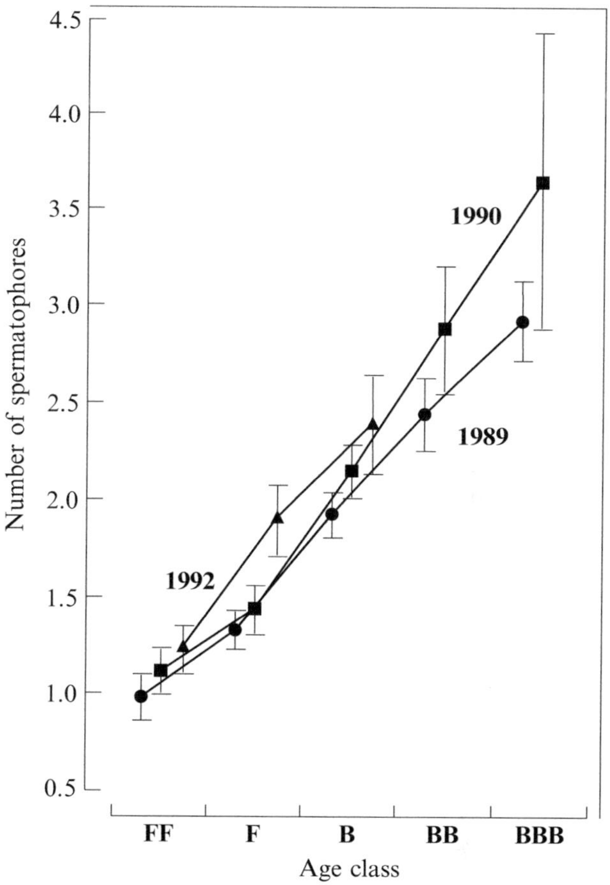

Fig. 2.15 Change in the mean number of spermatophores in wild-caught females of the cabbage white, *Pieris rapae*, for five age classes. *Circles*, *squares* and *triangles* indicate data for 1989, 1990 and 1992, respectively. Each *bar* represents SE. (From Watanabe and Ando 1993)

(Watanabe and Ando 1993). As the estimated male density changed from year to year in the habitat, as well as the fluctuation of weather conditions, the daily number of active males searching for mates must differ during the 3 years. Therefore, a species-specific lifetime mating number of females might be proposed, indicating that the lifetime mating number might be controlled by the female herself, not by the population density, male activities or other environmental factors.

Discussing the lifetime mating number of females seems to refer to a females-win scenario, wherein further increase in male defensive ability is checked either by a lack of additive genetic variation or by direct/indirect natural selection acting upon male ejaculate expenditure (Välimäki and Kaitala 2010). A mating number close to the female optimum must follow, if both the optimal female mating number and its potential to evolve are sufficiently high, and if the selection for increasing spermatophore depletion ability is intense enough. A three-way conflict among female reproductive physiology, male defence and male offence has been considered for the explanation of control of the female mating number.

References

Alcock J (2010) The hilltopping mating system of the duskywing skipper *Erynnis tristis* (Lepidoptera: Hesperiidae). J Res Lepid 43:1–5

Ando S, Watanabe M (1992) Egg load and multiple matings of a cabbage butterfly, *Pieris canidia indica* (Evans), in the wild. Jpn J Appl Entomol Zool 36:200–201 (in Japanese with English summary)

Ando S, Watanabe M (1993) Mating frequency and egg load in the white butterfly, *Pieris melete* Menetries, in a wild environment. Jpn J Ecol 43:111–114 (in Japanese)

Bergman M, Wiklund C (2009) Visual mate detection and mate flight pursuit in relation to sunspot size in a woodland territorial butterfly. Anim Behav 78:14–23

Birch MC, Lucas D, White PR (1989) The courtship behavior of the cabbage moth, *Mamestra brassicae* (Lepidoptera: Noctidae), and the role of male hair-pencils. J Insect Behav 2:227–239

Bissoondath CJ, Wiklund C (1995) Protein content of spermatophores in relation to monandry/polyandry in butterflies. Bahav Ecol Sociobiol 37:365–371

Bissoondath CJ, Wiklund C (1997) Effect of male body size on sperm precedence in the polyandrous butterfly *Pieris napi* L. (Lepidoptera: Pieridae). Behav Ecol 8:518–523

Boggs CL, Gilbert LE (1979) Male contribution to egg production in butterflies: evidence for transfer of nutrients at mating. Science 206:83–84

Boggs CL, Nieminen M (2004) Checkerspot reproductive biology. In: Ehrlich PR, Hanski I (eds) On the wings of checkerspots. Oxford University Press, Oxford, pp 92–111

Bulmer MG (1983) Models for the evolution for protandry in insects. Theor Popul Biol 23:314–322

Burns JM (1968) Mating frequency in natural populations of skippers and butterflies as determined by spermatophore counts. Proc Natl Acad Sci USA 61:852–859

Caballero-Mendieta N, Cordero C (2013) Male mating costs in a butterfly that produces small ejaculates. Physiol Entomol 38:318–325

Crudgington HS, Siva-Jothy MT (2000) Genital damage, kicking and early death. Nature (Lond) 407:855–856

Deinert EI, Lobgino JT, Gilbert LE (1994) Mate competition in butterflies. Nature (Lond) 370:23–24

Ehrlich PR (1965) The population biology of the butterfly *Euphydryas editha*. II. The structure of the Jasper Ridge colony. Evolution 19:327–336

Ehrlich PR, Gilbert LE (1973) Population structure and dynamics of the tropical butterfly *Heliconius ethilla*. Biotropica 5:69–82

Estrada C, Gilbert LE (2010) Host plants and immatures as mate-searching cues in *Heliconius* butterflies. Anim Behav 80:231–239

Fagerstrom T, Wiklund C (1982) Why do males emerge before females? Protandry as a mating strategy in male and female butterflies. Oecologia (Berl) 52:164–166

Fincke OM (1982) Lifetime mating success in a natural population of the damselfly, *Enallagma hageni* (Walsh) (Odonata: Coenagrionidae). Bahav Ecol Sociobiol 10:293–302

Fiske P, Rintamäki PT, Karvonen E (1998) Mating success in lekking males: a meta-analysis. Behav Ecol 9:328–338

Forsberg J, Wiklund C (1988) Protandry in the green-veined white butterfly, *Pieris napi* L. (Lepidoptera; Pieridae). Funct Ecol 2:81–88

Fosberg J, Wiklund C (1989) Mating in the afternoon: time-saving in courtship and remating by females of the polyandrous butterfly *Pieris napi* L. Behav Ecol Sociobiol 25:349–356

Grula JW, McChesney JD, Taylor OR Jr (1980) Aphrodisiac pheromones of the sulfur butterflies *Colias eurytheme* and *C. philodice* (Lepidoptera, Pieridae). J Chem Ecol 6:241–256

Hasegawa M, Watanabe M (2008) Changes in mate preference of male *Colias erate poliographus* in relation to copulation experience of females. Jpn J Entomol 11:105–114 (in Japanese with English summary)

Irie S, Watanabe M (2009) Courtship behavior of the male sulfur butterfly *Colias erate* toward experienced females that were inhibited from showing mate refusal behavior by experimental manipulation. Jpn J Biol Educ 49:68–75 (in Japanese with English summary)

Iwasa Y, Odendaal FJ, Murphy DD, Ehrlich PR, Launer AE (1983) Emergence patterns in male butterflies: a hypothesis and a test. Theor Popul Biol 23:363–379

Kaitala A, Wiklund C (1995) Female mate choice and mating costs in the polyandrous butterfly *Pieris napi* (Lepidoptera: Pieridae). J Insect Behav 8:355–363

Karlsson B (1996) Male reproductive reserves in relation to mating system in butterflies: a comparative study. Proc R Soc Lond B 263:187–192

Kemp DJ, Wiklund C (2001) Fighting without weaponry: a review of male–male contest competition in butterflies. Behav Ecol Sociobiol 49:419–442

Kemp DJ, Macedonia JM, Ball TS, Rutowski RL (2008) Potential direct fitness consequences of ornament-based mate choice in a butterfly. Behav Ecol Sociobiol 62:1017–1026

Kiritani K, Nozato K, Miyai S, Ban Y (1984) Studies on ecology and behavior of Japanese black swallowtail butterflies. I. Ecological characteristics of male populations in *Papilio helenus nicconicolens* Butler and *P. protenor demetrius* Cramer (Lepidoptera: Papilionidae). Appl Entomol Zool 19:509–517

Klein AL, de Araujo AM (2010) Courtship behavior of *Heliconius erato phyllis* (Lepidoptera, Nymphalidae) towards virgin and mated females: conflict between attraction and repulsion signals? J Ethol 28:409–420

Konagaya T, Watanabe M (2015) Adaptive significance of the mating of autumn-morph females with non-overwintering summer-morph males in the Japanese common grass yellow, *Eurema mandarina* (Lepidoptera: Pieridae). Appl Entomol Zool 50:41–47

Konagaya T, Mutoh N, Suzuki M, Rutowski RL, Watanabe M (2015) Estimates of female lifetime fecundity and changes in the number and types of sperm stored with age and time since mating in the monandrous swallowtail butterfly, *Battus philenor* (Lepidoptera: Papilionidae) in the Arizona desert. Appl Entomol Zool 50:311–316

Labine PA (1968) The population biology of the butterfly, *Euphydryas editha*. VIII. Oviposition and its relation to patterns of oviposition in other butterflies. Evolution 22:799–805

LaMunyon CW, Eisner T (1994) Spermatophore size as determinant of paternity in an arctiid moth (*Utetheisa ornatrix*). Proc Natl Acad Sci USA 91:7081–7084

Lederhouse RC (1981) The effect of female mating frequency on egg fertility in the black swallowtail, *Papilio polyxenes asterius* (Papilionidae). J Lepid Soc 35:266–277

Leopold RA (1976) The role of male accessory glands in insect reproduction. Annu Rev Entomol 21:199–221

Marshall LD (1985) Protein and lipid composition of *Colias philodice* and *C. eurytheme* spermatophores and their changes over time. J Res Lepid 24:21–30

Nakanishi Y, Watanabe M, Ito T (1996) Differences in lifetime reproductive output and mating frequency of two female morphs of the sulfur butterfly, *Colias erate* (Lepidoptera: Pieridae). J Res Lepid 35:1–8

Nozato K, Kiritani K, Miyai S, Ban Y (1985) Studies on ecology and behavior of Japanese black swallowtail butterflies. IV. Estimation of life span of male *Papilio helenus nicconicolens* Butler and *P. protenor demetrius* Cramer (Lepidoptera: Papilionidae). Appl Entomol Zool 20:494–496

Osanai M, Aigaki T, Kasuga H (1987) Energy metabolism in the spermatophore of the silkmoth, *Bombyx mori*, associated with accumulation of alanine derived from arginine. Insect Biochem 17:71–75

Parker GA (1970) Sperm competition and its evolutionary consequences in insects. Biol Rev 45:525–567

Parker GA (1982) Why are there so many tiny sperm? Sperm competition and the maintenance of two sexes. J Theor Biol 96:281–294

Rajyaguru P, Pegram KV, Kingston AC, Rutowski RL (2013) Male wing color properties predict the size of nuptial gifts given during mating in the pipevine swallowtail butterfly (*Battus philenor*). Naturwissenschaften 100:507–513
Rosenberg RH, Enquist M (1991) Contest behaviour in Weidemeyer's admiral butterfly *Limenitis weidemeyerii* (Nymphalidae): the effect of size and residency. Anim Behav 42:805–811
Rutowski RL (1978) The courtship behavior of the small sulphur butterfly, *Eurema lisa* (Lepidoptera: Pieridae). Anim Behav 26:892–903
Rutowski RL (1984) Sexual selection and the evolution of butterfly mating behavior. J Res Lepid 23:125–142
Rutowski RL (1991) The evolution of male mate-locating behavior in butterflies. Am Nat 138:1121–1139
Rutowski RL, Gilchrist GW (1986) Copulation in *Colias eurytheme* (Lepidoptera: Pieridae): patterns and frequency. J Zool Lond A 209:115–124
Rutowski RL, Rajayaguru P (2013) Male-specific iridescent coloration in the pipevine swallowtail (*Battus philenor*): used in mate choice but not sexual discrimination by males. J Insect Behav 26:200–211
Rutowski RL, Newton M, Schaefer J (1983) Interspecific variation in the size of the nutrient investment made by male butterflies during copulation. Evolution 37:708–713
Rutowski RL, Alcock J, Carey M (1989) Hiltopping in the pipevine swallowtail butterfly (*Battus philenor*). Ethology 82:244–254
Rutowski RL, Nahm AC, Macedonia JM (2010) Iridescent hindwing patches in the pipevine swallowtail: differences in dorsal and ventral surface relate to signal function and context. Funct Ecol 24:767–775
Sasaki N, Konagaya T, Watanabe M, Rutowski RL (2015) Indicators of recent mating success in the pipevine swallowtail butterfly (*Battus philenor*) and their relationship to male phenotype. J Insect Physiol 83:30–36
Scott JA (1972) Mating of butterflies. J Res Lepid 11:99–127
Scott JA (1975) Variability of courtship of the buckeye butterfly, *Precis coenia* (Nymphalidae). J Res Lepid 14:142–147
Scriber JM, Lederhouse RC (1988) Hand-pairing of *Papilio glaucus glaucus* and *Papilio pilumnus* (Papilionidae) and hybrid survival on various food plants. J Res Lepid 27:96–103
Shreeve TG (1987) The mate location behaviour of the male speckled wood butterfly, *Pararge aegeria*, and the effect of phenotypic differences in hind-wing spotting. Anim Behav 35:682–690
Silberglied RE, Taylor OR Jr (1978) Ultraviolet reflection and its behavioral role in the courtship of sulfur butterflies *Colias eurytheme* and *C. philodice* (Lepidoptera, Pieridae). Behav Ecol Sociobiol 3:203–243
Singer MC (1982) Sexual selection for small size in male butterflies. Am Nat 119:440–443
Stern VM, Smith RF (1960) Factors affecting egg production and oviposition in populations of *Colias philodice eurytheme* Boisduval (Lepidoptera: Pieridae). Hilgardia 29:411–454
Sugawara T (1979) Stretch reception in the bursa copulatrix of the butterfly, *Pieris rapae crucivora*, and its role in behaviour. J Comp Physiol 130:191–199
Svärd L, Wiklund C (1988) Prolonged mating in the monarch butterfly *Danaus plexippus* and nightfall as a cue for sperm transfer. Oikos 52:351–354
Svärd L, Wiklund C (1989) Mass and production rate of ejaculates in relation to monandry/polyandry in butterflies. Behav Ecol Sociobiol 24:395–402
Takanashi T, Hiroki M, Obara Y (2001) Evidence for male and female sex pheromones in the sulfur butterfly, *Eurema hecabe*. Entomol Exp Appl 101:89–92
Tatar M (1991) Clutch size in the swallowtail butterfly, *Battus philenor*: the role of host quality and egg load within and among seasonal flights in California. Behav Ecol Sociobiol 28:337–344

Thibout E (1979) Stimulation of reproductive activity of females of *Acrolepiosis assectella* (Lepidoptera: Hyponomeutoidea) by the presence of eupyrene spermatozoa in the spermatheca. Entomol Exp Appl 26:279–290

Tschudi-Rein K, Benz G (1990) Mechanisms of sperm transfer in female *Pieris brassicae* (Lepidoptera: Pieridae). Ann Entomol Soc Am 83:1158–1164

Välimäki P, Kaitala A (2010) Properties of male ejaculates do not generate geographical variation in female mating tactics in a butterfly *Pieris napi*. Anim Behav 79:1173–1179

Vande Velde L, Damiens D, Van Dyck H (2011) Spermatophore and sperm allocation in males of the monandrous butterfly *Pararge aegeria*: the female's perspective. Ethology 117:645–654

Vane-Wright D (2003) Butterflies. The Natural History Museum, London

Watanabe M (1978) Adult movements and resident ratios of the black-veined white, *Aporia crataegi*, in a hilly region. Jpn J Ecol 28:101–109

Watanabe M (1979a) Natural mortalities of the swallowtail butterfly, *Papilio xuthus* L., at patchy habitats along the flyways in a hilly region. Jpn J Ecol 29:85–93

Watanabe M (1979b) Population sizes and resident ratios of the swallowtail butterfly, *Papilio polytes* L., at a secondary bush community in Dharan, Nepal. Kontyu 47:291–297

Watanabe M (1988) Multiple matings increase the fecundity of the yellow swallowtail butterfly, *Papilio xuthus* L., in summer generations. J Insect Behav 1:17–29

Watanabe M, Ando S (1993) Influence of mating frequency of lifetime fecundity in wild females of the small white *Pieris rapae* (Lepidoptera, Pieridae). Jpn J Entomol 61:691–696

Watanabe M, Hirota M (1999) Effects of sucrose intake on spermatophore mass produced by male swallowtail butterfly *Papilio xuthus* L. Zool Sci 16:55–61

Watanabe M, Imoto T (2003) Thermoregulation and flying habits of the Japanese sulfur butterfly *Colias erate* (Lepidoptera: Pieridae) in an open habitat. Entomol Sci 6:111–118

Watanabe M, Irie S (2011) Female abdomen position signals receptivity during courtship in the Japanese sulphur butterfly, *Colias erate* (Pieridae). J Lepid Soc 65:259–263

Watanabe M, Nishimura M (2001) Reproductive output and egg maturation in relation to mate-avoidance in monandrous females of the small copper, *Lycaena phlaeas* (Lycaenidae). J Lepid Soc 54:83–87

Watanabe M, Nozato K (1986) Fecundity of the yellow swallowtail butterflies, *Papilio xuthus* and *P. machaon hippocrates*, in a wild environment. Zool Sci 3:509–516

Watanabe M, Sasaki N (2010) Pattern of sperm storage and migration in the reproductive tract of the swallowtail butterfly *Papilio xuthus*: cryptic female choice after second mating. Physiol Entomol 35:328–333

Watanabe M, Sato K (1993) A spermatophore structured in the bursa copulatrix of the small white *Pieris rapae* (Lepidoptera, Pieridae) during copulation and its sugar content. J Res Lepid 32:26–36

Watanabe M, Nozato K, Kiritani K (1986) Studies on ecology and behavior of Japanese black swallowtail butterflies (Lepidoptera: Papilionidae). 5. Fecundity in summer generations. Appl Entomol Zool 21:448–453

Watanabe M, Nakanishi Y, Bon'no M (1997) Prolonged copulation and spermatophore size ejaculated in the sulfur butterfly, *Colias erate* (Lepidoptera: Pieridae) under selective harassments of mated pairs by conspecific lone males. J Ethol 15:45–54

Watanabe M, Wiklund C, Bon'no M (1998a) Ejaculation timing of eupyrene and apyrene sperm in the cabbage white butterfly *Pieris rapae* (Lepidoptera, Pieridae) during copulation. Entomol Sci 1:15–19

Watanabe M, Wiklund C, Bon'no M (1998b) The effect of repeated matings on sperm numbers in successive ejaculates of the cabbage white butterfly *Pieris rapae* (Lepidoptera: Pieridae). J Insect Behav 11:559–570

Watanabe M, Bon'no M, Hachisuka A (2000) Eupyrene sperm migrates to spermatheca after apyrene sperm in the swallowtail butterfly, *Papilio xuthus* L. (Lepidoptera: Papilionidae). J Ethol 18:91–99

West DA (1983) Hand-pairing of *Battus philenor* (Papilionidae). J Lepid Soc 37:90

Wickman P-O (1985) Male determined mating duration in butterflies? J Lepid Soc 39:341–342
Wickman P-O (1986) Courtship solicitation by females of the small heath butterfly, *Coenonympha pamphilus* (L.) (Lepidoptera: Satyridae) and their behaviour in relation to male territories before and after copulation. Anim Behav 34:153–157
Wickman P-O, Wiklund C (1983) Territorial defence and its seasonal decline in the speckled wood butterfly (*Pararge aegeria*). Anim Behav 31:1206–1216
Wiklund C (1977) Courtship behavior in relation to female monogamy in *Leptidea sinapis* (Lepidoptera). Oikos 29:275–283
Wiklund C (1982) Behavioral shift from courtship solicitation to mate avoidance in female ringlet butterflies (*Aphantopus hyperanthus*) after copulation. Anim Behav 30:790–793
Wiklund C (2003) Sexual selection and the evolution of butterfly mating systems. In: Boggs CL, Watt WB, Ehrlich PR (eds) Butterflies: ecology and evolution taking a flight. University of Chicago Press, Chicago, pp 67–90
Wiklund C, Fagerstrom T (1977) Why do males emerge before females? a hypothesis to explain the incidence of protandry in butterflies. Oecologia (Berl) 31:153–158
Wiklund C, Forsberg J (1991) Sexual size dimorphism in relation to female polygamy and protandry on butterflies: a comparative study of Swedish Pieridae and Satyridae. Oikos 60:373–381
Wiklund C, Gotthard K, Nylin S (2003) Mating system and the evolution of sex-specific mortality rates in two nymphalid butterflies. Proc R Soc Lond B 270:1823–1828
Zonneveld C (1992) Polyandry and protandry in butterflies. Bull Math Biol 54:957–9765

Chapter 3
Reproductive Anatomy

Keywords Bursa copulatrix • Lagena • Oviposition • Signa • Somatic maintenance • Spermatheca • Spermathecal gland • Spermatophore • Testes

The reproductive systems of males and females are generally constant in form within butterfly species. Males provide sperm as well as accessory gland substances for transferring to females. During copulation, a single spermatophore including sperm is structured in the bursa copulatrix of females. The bursa copulatrix can contain a full-size spermatophore or remnants of several spermatophores. The sperm to be used for fertilisation is stored in the spermatheca of females, not in the spermatophore. The female oviposition system is essentially a paired structure with two ovaries. Fertilisation takes place in the common oviduct as the sperm pass along this duct and enter the egg. Sperm dimorphism occurs in butterflies. An eupyrene spermatozoon, the normal sperm that can fertilise an egg, is transferred to the female as bundles in the spermatophore. A bundle consists of 256 eupyrene spermatozoa. The apyrene spermatozoon, however, has no nucleus and does not have fertilisation ability. After copulation, both types of spermatozoa move out of the bursa copulatrix, and then arrive at the spermatheca to be stored until oviposition. The spermatheca does not enlarge relative to the stored spermatozoa, suggesting the number of spermatozoa stored is limited. Therefore, the success of sperm migration from the spermatophore to the spermatheca is primarily critical for the reproductive success of the males.

Reproductive Organs in Males and Females

A diagram of the internal reproductive organs in male butterflies is shown in Fig. 3.1. In most butterflies, the two testes are fused and give rise to the paired vas deferens, tubular secretory ducts that lead to the duplex, which is a pair of sperm storage organs that unite caudally to form the simplex, a single duct leading to the intromittent organ or aedeagus. In general, the simplex of a virgin male is a long tube, often filled with a white, granular substance near the aedeagus and with a colourless pasty substance in the rest of the organ. Sperm move from the testes to the duplex via the vas deferens with a daily rhythmic release (Riemann et al. 1974).

M. Watanabe, *Sperm Competition in Butterflies*, Ecological Research Monographs,
DOI 10.1007/978-4-431-55945-0_3

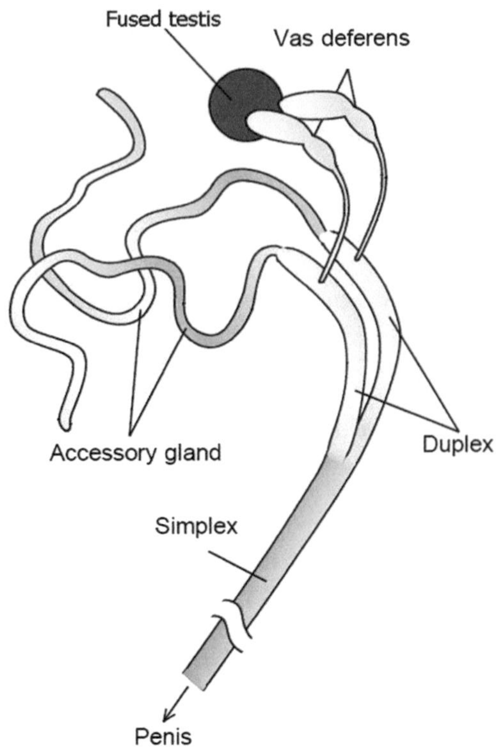

Fig. 3.1 Schematic representation of the reproductive organs of male swallowtail butterflies (From Sasaki et al. 2015)

Hiroyoshi and Mitsuhashi (1999) showed that some sperm in the duplex move back to the vas deferens during mating, and that such sperm reflex provides a means to save sperm. This observation suggests that males have the ability to regulate the number of sperm transferred. Conversely, the number of sperm in the duplex must also vary, even among males just after mating. Indeed, it has been shown that the number of sperm transferred by males was affected by various factors such as the male mating history (Watanabe et al. 1998), female mating history (Cook and Gage 1995) and relatedness (Lewis and Wedell 2009). Because sperm remained in the duplex of the male after mating, the number of sperm found in the duplex is not a good indicator of male mating history.

Because of the arrangement of the reproductive organs in the male body, the spermatophore materials and the accessory substance in the simplex flow into the female body throughout copulation, followed by the sperm (Watanabe and Sato 1993). In the swallowtail butterfly, *Papilio xuthus*, the spermatophore and accessory substances are transferred to the bursa copulatrix of the female separately, first accessory substances and then the spermatophore (Watanabe et al. 2000). Sperm must be transferred to the female after the transfer of this clear pasty substance. Thus, males might not be able to retain the spermatophore materials and the accessory substances in the simplex when ejaculation starts. Then, after copulation, components of the male reproductive tracts would become shorter, decrease in mass, and change in colour. Sasaki et al. (2015) reported that these changes persisted at least for 2 days after mating in the pipevine swallowtail, *Battus philenor*.

Brief copula duration must be of potential direct benefit for both sexes of butterflies, because mating pairs are probably exposed to a greater risk of predation when relatively inactive. However, females have to wait until copulation termination. Because the sperm are transferred after the transfer of spermatophore materials and the accessory substances (Watanabe and Sato 1993), according to the male reproductive morphology, the male needs a certain period of copulation time to transfer enough sperm. Thus, males should control the duration of the copulation (Wickman 1985).

Two reproductive organs in the female of many butterfly species, that is, the bursa copulatrix, which is a receptacle for spermatophores, and the spermatheca, which is a storage for sperm, has been described in terms of sperm competition for fertilisation of the eggs. The layout of the reproductive organs with several glands in the female is shown in Fig. 3.2. The glands shown venting into the common oviduct are of largely unknown function, but it is thought that the large convoluted pairs of glands provide the adhesive which cements eggs to leaves (Porter 1992). During a mating, a male butterfly transfers a single spermatophore which contains sperm, providing sufficient sperm to fertilise all the eggs produced. The bursa copulatrix, initially a thin, flat ellipse, grows rapidly in volume and becomes oval at about the time the spermatophore is structured. In the cabbage white, *Pieris rapae*, muscles of the bursa copulatrix extend radially from the sugnum area (ventral) to an exposed dorsal epithelial area (Sugawara 1979).

During copulation, male secretions from the ductus ejaculatorius and accessory substances seems to be transferred serially to the bursa copulatrix. The ejaculate is a white gel mainly observed on the tip of the penis 20 min after the beginning of copulation in the swallowtail butterfly, *Papilio xuthus* (Watanabe et al. 2000). No spermatophore capsule in the bursa copulatrix is structured until the copulation has lasted at least 30 min. Just after the termination of copulation (~1 h in many species), the spermatophore represents a balloon-like structure with a swollen corpus and a tube-like column with an opening at its tip.

The luminal surface of the bursa copulatrix wall may be more or less smooth, but frequently it is variably covered by microprotuberances, which may give a dentate, scobinate or rough appearance (Lincango et al. 2013). To release the contents of the spermatophore such as sperm and seminal fluid, the female has to rupture it; this is usually accomplished by special teeth, the signa, on the inside of the bursa copulatrix (Fig. 3.3). These teeth mechanically puncture the spermatophore when females contract the muscles surrounding the bursa copulatrix (Fig. 3.4). Then, the sperm that flowed in the bursa copulatrix begin to migrate towards the spermatheca, via the duct seminalis and the spermathecal duct, which might exhibit a peristaltic movement. Therefore, the signa might have evolved as a means whereby females can influence the rate of spermatophore digestion and hence their receptivity to the next mating. Figure 3.5 shows a spermatophore collapsed by the signa of a bursa copulatrix compared with an intact spermatophore just after termination of

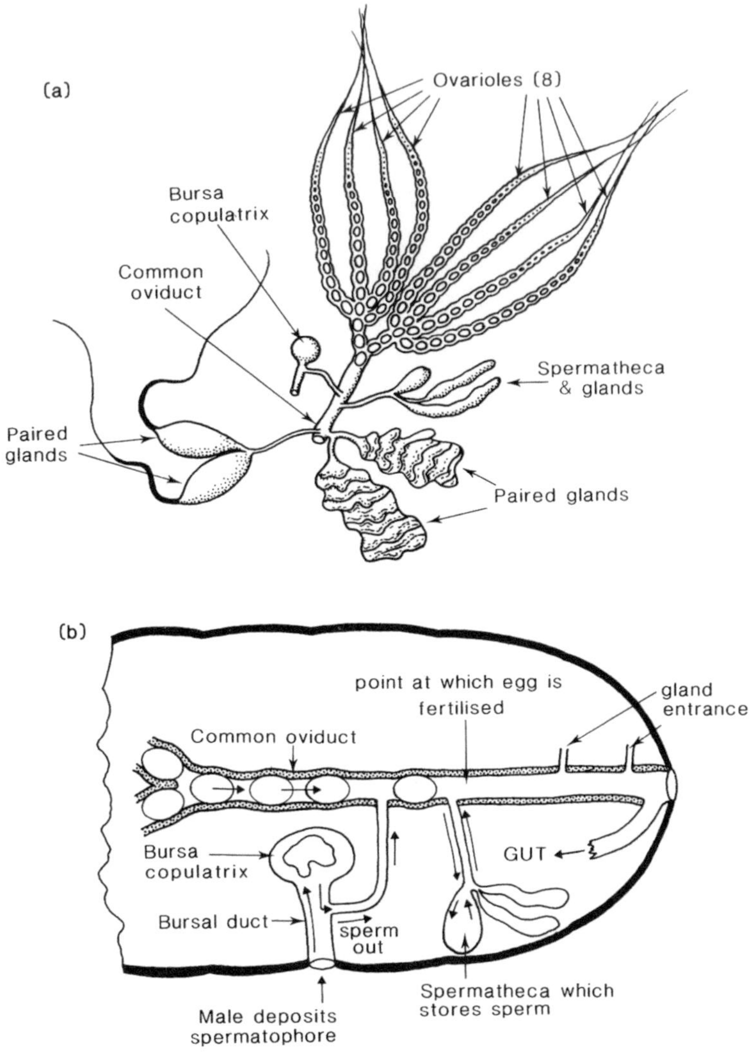

Fig. 3.2 Layout of the female reproductive system of the marsh fritillary, *Eurodryas aurinia* (From Porter 1992). (**a**) Reproductive organs. (**b**) Cross-sectional view

copulation in the swallowtail butterfly, *Papilio xuthus*. Galicia et al. (2008) considered an additional function of the signa, such as holding and detecting the spermatophore or protecting the bursa integument, of the seven hypotheses that have been proposed on the function of the signa in lepidopteran females:

1. Organ for locking with complementary structures of the penis
2. Organ for stimulating ejaculation in males
3. Organ for protecting the corpus bursa from damage inflicted by penile structures

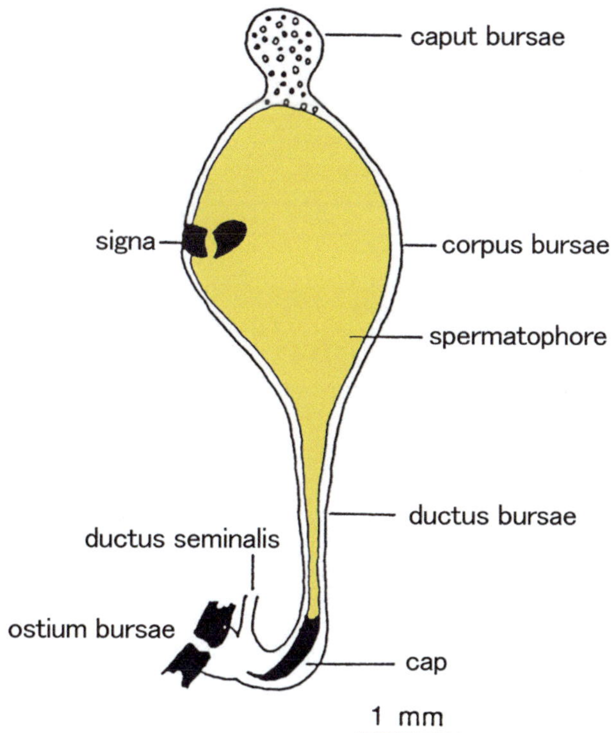

Fig. 3.3 The bursa copulatrix with a spermatophore soon after copulation in the large white, *Pieris brassicae*. The orifice of the spermatophore is sealed with a cap produced by secretion of the distal portion of the male accessory glands (From Tschudi-Rein and Benz 1990)

4. Sense organ for detecting the presence, and/or assessing the size, of the spermatophore received during copulation
5. Organ for holding the spermatophores
6. Organ for breaking (by puncturing, tearing and/or sectioning) the spermatophore capsule;
7. Organ for filtering particles of degraded spermatophore that could block the ductus seminalis, preventing sperm migration to the spermatheca

Furthermore, Sánchez et al. (2011) pointed out that polyandrous species would have developed a signum and thicker spermatophore walls, whereas monandrous species would favour loss of signa and thinner walls.

After the termination of copulation, sperm have to migrate from the spermatophore to the spermatheca for storing and fertilisation at oviposition, via the seminal duct and the spermathecal duct (Fig. 3.6). In the swallowtail butterfly, *Papilio xuthus*, the length of the former is 3.54 mm and that of the latter is 1.64 mm (Watanabe et al. 2000). Therefore, the sperm have to move more than 5.2 mm during sperm migration, suggesting high mortality.

The spermatheca has been the designated organ at which sperm has to arrive from the spermatophore until fertilisation. In addition to the main chamber of the spermatheca, sperm are commonly found in the smaller chamber, the lagena, and in

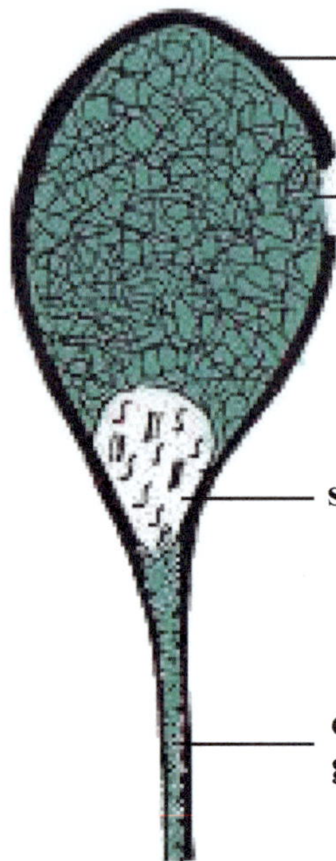

Fig. 3.4 The spermatophore about 2 h after copulation in the large white, *Pieris brassicae*. The cap around the orifice has disintegrated, and the imprint of the signa is visible in the wall of the corpus. The wall of the spermatophore, as well as the white secretion within the corpus, are derived from secretions of the male ejaculatory duct; the yellow granular secretion filling the spermatophore column comes from the proximal portion of the male accessory glands (From Tschudi-Rein and Benz 1990)

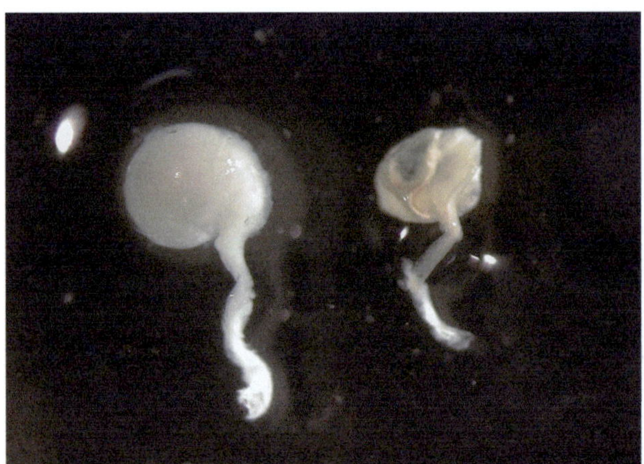

Fig. 3.5 An intact spermatophore just after termination of copulation (*left*) and a collapsed spermatophore a few days after the copulation (*right*) in the swallowtail butterfly, *Papilio xuthus*

the seminal duct. Watanabe et al. (2000) described the morphology of the spermathecal gland leading from the distal end of the spermatheca of the swallowtail butterfly, *Papilio xuthus*. The length and width of the spermatheca are 2.8 mm and 0.4 mm, respectively. The spermathecal gland is 31.0 mm, with a diameter of 0.2 mm (Fig. 3.7). Probably the spermathecal gland does not change in size

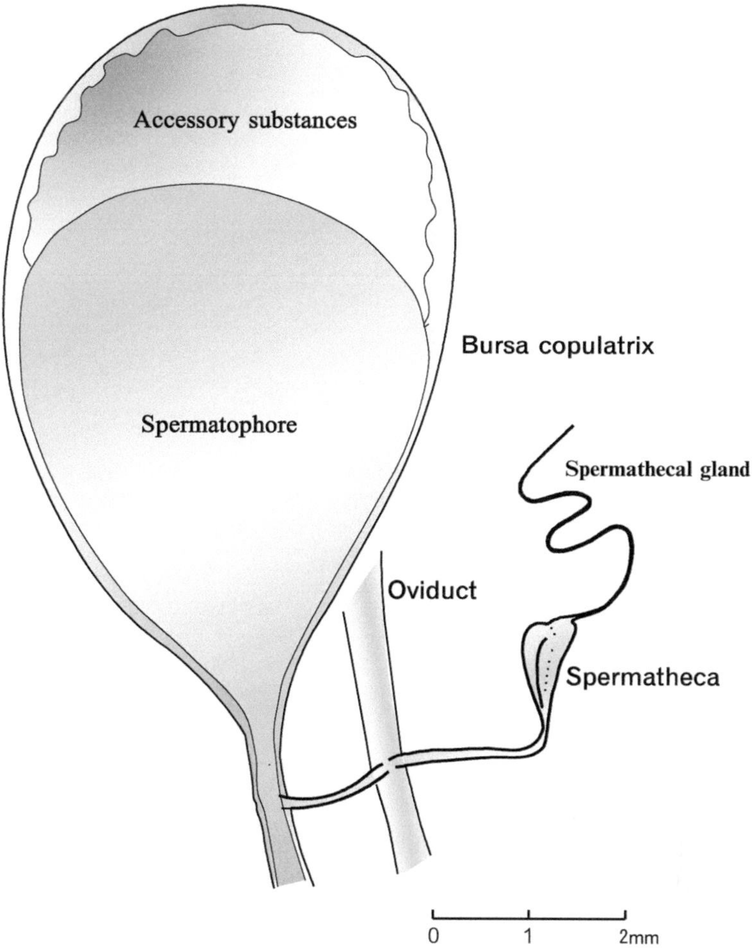

Fig. 3.6 Female reproductive organs in the swallowtail butterfly, *Papilio xuthus*. Drawing made from dissection of a just-mated female. A single spermatophore is deposited in the bursa copulatrix. Sperm migrate from the bursa copulatrix through the seminal duct to the oviduct, and then enter the spermathecal duct on their way to the spermatheca, to which the spermathecal gland is attached. The ovarioles supply unfertilised mature eggs

irrespective of amount of sperm contained. Drummond (1984) stated that the gland supplies nutrient to keep the spermatozoa alive and that some spermatozoa exist in it. He et al. (1995) showed that secretion by the gland of the oriental armyworm, *Pseudaletia separata*, activated spermatozoa. In the potato moth, *Phthorimaea operculella*, the gland secretes fluid which keeps sperm alive in the spermatheca and also assists in the movement of eupyrene sperm to the site of fertilisation (Nabi and Harrison 1983). Although morphology of the gland varies widely among butterfly species (Fig. 3.8), all are a cul-de-sac tube providing possible sperm storage. Therefore, in the female swallowtail butterfly the destination of sperm

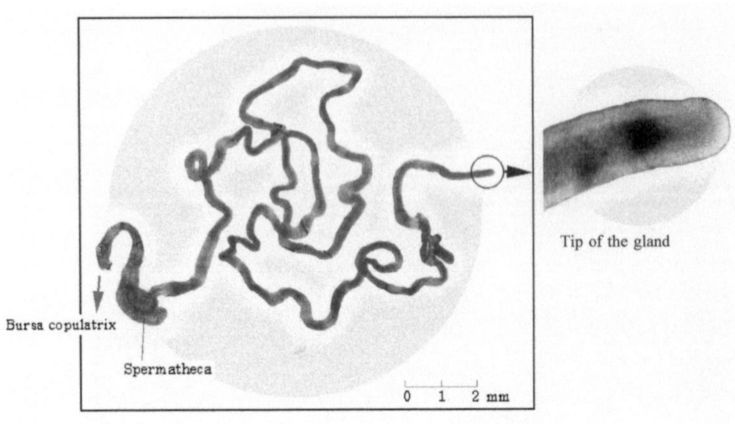

Fig. 3.7 A spermathecal gland attached to the spermatheca under stereoscopic microscope and an enlarged photograph of the tip in the swallowtail butterfly, *Papilio xuthus* (From Watanabe and Hachisuka 2005)

migrating from the spermatophore is the tip of the gland via the spermatheca (Watanabe and Hachisuka 2005).

Fecundity

When a butterfly emerges from the pupa, it has a reserve of energy stored in the fat body of the abdomen, a limited reserve of nitrogen, and some trace elements, although the actual amounts are dependent on how favourable conditions were in the larval stage. During dissection of the abdomen for examining the reproductive organs of female butterflies, a considerable changes in the quantity of fat body and the air sac volume with the female age might be found when removing the cuticle. As shown in Fig. 3.9, the newly emerged or fresh females (age FF) of the swallowtail butterfly, *Papilio xuthus*, have the most abundant fat bodies, which filled the cavity of the abdomen, preventing observing any reproductive organs as well as any digestive organs (Watanabe and Nozato 1986). Thus, the fat bodies have to be carefully removed to dissect out the reproductive organs (Fig. 3.10). On the other hand, the oldest females (age BBB) have the smallest or a little fat body with the largest air sac. Matured eggs in the ovaries could be easily observed.

The female reproductive system for oviposition is essentially a paired structure with two ovaries, each ovary comprising four ovarioles. Ovaries of the female initially contain previtellogenic oocytes, and vitellogenesis begins at eclosion. Eight ovarioles are coiled in the posterior fat body tissue of the female. Each ovariole consists of a distal germarium in which new oocytes arise, following proximally by a graded series of enlarging oocytes. Eggs begin their development

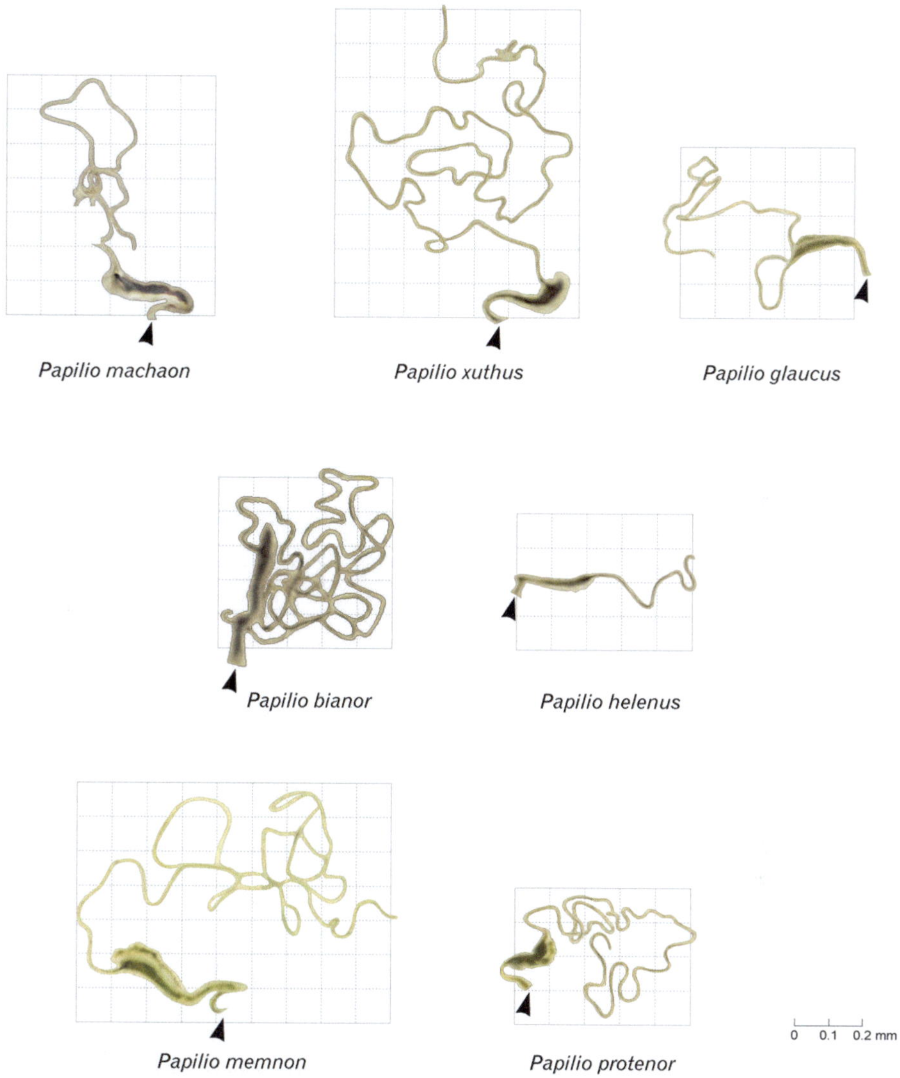

Fig. 3.8 Spermatheca with spermathecal gland in swallowtail butterflies. Each *arrow* represents a connection of the spermathecal duct

at the tips of the ovarioles and move along the increasingly wider tube as they grow in size. Each egg is surrounded by seven nurse cells, which absorb nutrients from the female's haemolymph and pass it into the growing egg. The nurse cells eventually form a membrane (follicle) which surrounds the mature egg, and this is shed as the eggs pass into the common oviduct. The follicle cells secrete the materials that build the eggshell (Porter 1992).

Fig. 3.9 Photographs of the abdomens of the swallowtail butterfly, *Papilio xuthus*, with the cuticle removed. Abundant fat bodies without an air sac prevent observing the eggs in age FF; several mature eggs can be seen in age BBB because there are fewer fat bodies with a large air sac (From Watanabe and Nozato 1986)

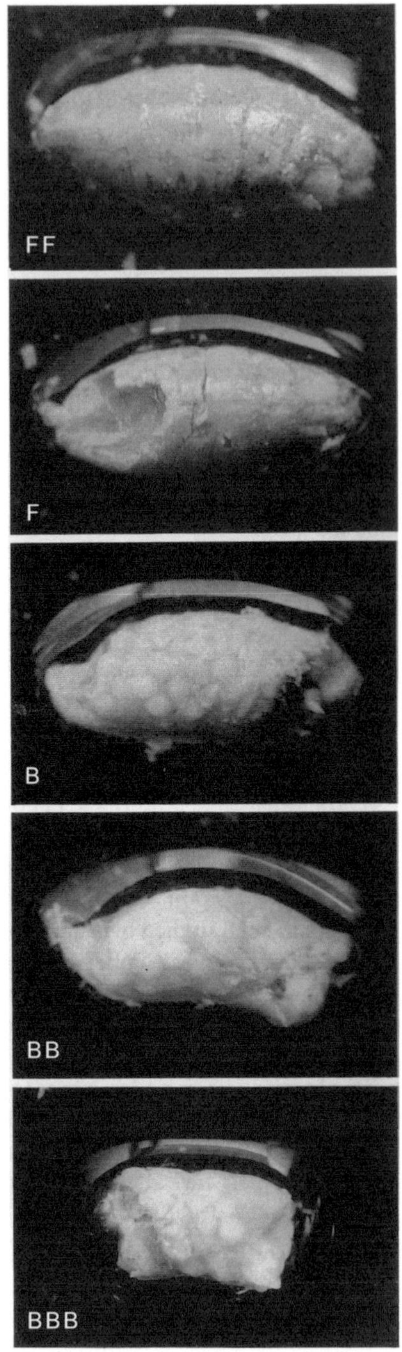

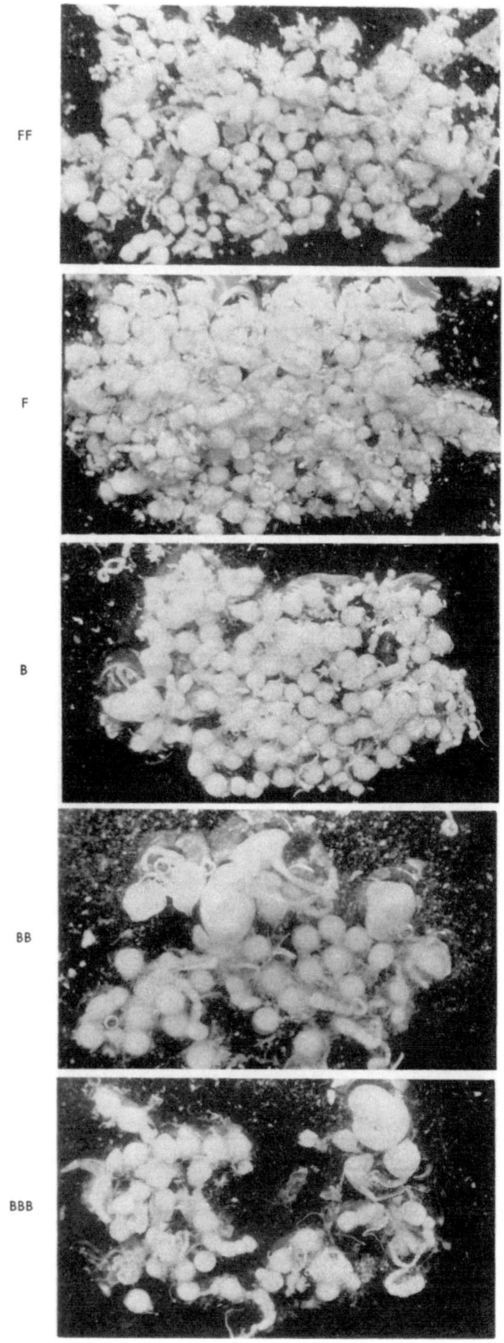

Fig. 3.10 Eggs and spermatophores isolated from abdomen of female swallowtail butterfly, *Papilio xuthus*

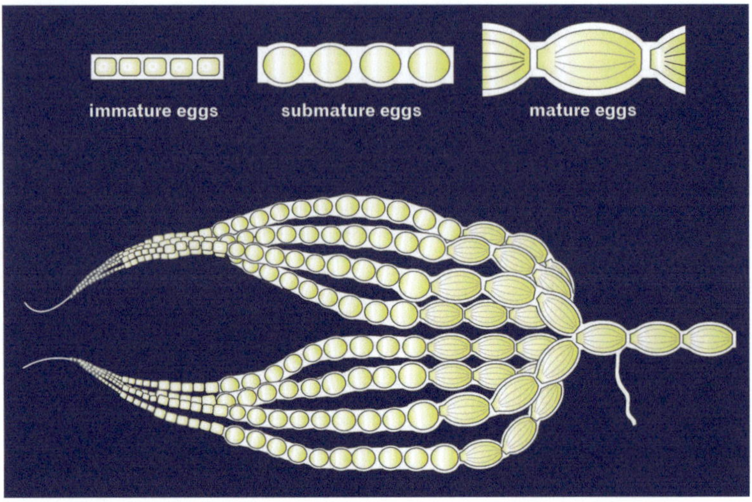

Fig. 3.11 Schematic representation of eggs in ovaries of female pierid butterfly

Dunlap-Pianka et al. (1977) describe three major stages in oogenesis, as follows:

1. Previtellogenesis, primarily a period of nurse cell development
2. Vitellogenesis, entailing rapid oocyte growth through accumulation of yolk, and eventual nurse cell shrinkage
3. Chorion secretion by the follicle cells

Then, the mature eggs ovulate from the follicular envelope into one of the eight branches of the oviduct (Fig. 3.11). Therefore, by dissection of the female, various shapes and size of eggs in the ovaries could be observed and be classified into three groups: mature eggs, submature eggs and immature eggs, as reported by Watanabe and Ando (1993). A mature egg has a hard eggshell, to be sufficient for the size of the egg laid. A submature egg is filled with yolk but without an eggshell. An immature egg includes oocytes with little yolk. In general, the number of mature eggs and submature eggs could be counted directly under the stereoscopic microscope (×30), whereas the number of immature eggs should be estimated from the number of immature eggs per ovariole and the number of ovarioles under the stereoscopic microscope (×75).

Because females exhibit minimal parental concern for their offspring, the best thing females can do for their young is to place the eggs on the correct available host plants. Consequently, the newly hatched caterpillars can easily locate their substrate and begin to feed. However, each life table of the egg stage in butterfly species shows considerable mortality rates by egg parasitic wasps as well as predatory mites, spiders, ants and other predators (Watanabe 1981). With such a high mortality in the egg stage, butterflies seem to have evolved a strategy of producing many eggs. Feltwell (1986) stated that some species are supposed to lay more than 500 eggs but the average number actually laid is thought to be between 200 and 300.

Just after eclosion of the females in the cabbage white, *Pieris rapae*, such virgin females loaded about 450 immature eggs with relatively small number of mature and submature eggs; thus, Watanabe and Ando (1993) concluded that the fecundity is 550. By dissection of the young females (= FF stage) of the swallowtail butterflies, *Papilio xuthus* and *Papilio machaon*, in the field, fecundity is estimated at 500 to 600, irrespective of generations and species (Watanabe and Nozato 1986). Changes in the number of eggs in the ovaries with age class of both species are shown in Table 3.1. The numbers of submature, immature and total eggs decreased with age, with the exception of the summer generation of *P. machaon*. However, the number of mature eggs seems not to decrease with age in all the generations. In *P. xuthus*, the number of mature eggs in ages FF to BB of the spring generation are smaller than those of the summer generation. However, the numbers of submature and immature eggs in ages FF to B are not significantly different between the two generations, although those in age BB and BBB in the spring generation are smaller than those in the summer generation. Therefore, the fecundity of both spring and summer generations is not so different, indicating that the small body of a spring generation adult might affect not the fecundity but the mature egg size in the ovaries. That is, the diameter of eggs laid by the spring and summer generation females is about 1.15 mm and 1.20 mm, respectively.

For the black swallowtail butterflies, *Papilio helenus* and *Papilio protenor*, the estimated fecundity was 400 to 500 (Watanabe et al. 1986). The fecundity of field-captured virgin females, assumed to be recently emerged, in the Asian green-veined white, *Pieris melete*, the Indian cabbage white, *Pieris canidia*, and the eastern pale clouded yellow, *Colias erate*, is 350 (Ando and Watanabe 1993), 300 (Ando and Watanabe 1992) and 800 (Nakanishi et al. 1996), respectively.

In the small copper, *Lycaena phlaeas*, which is a monandrous species, the mean number of eggs in wild-caught virgin females of age FF is 42, 19 and 242 for mature, submature and immature eggs, respectively. The mated females of age FF contained 261 immature eggs, a number similar to that for virgin females. Therefore, the potential fecundity of the small copper is estimated at about 300 (Watanabe and Nishimura 2001). Figure 3.12 shows the lifetime changes in the number of immature eggs in monogamous females, a number that decreases with age. Because there are fewer than 20 mature eggs in the ovaries of females of age BBB, the decrease in immature eggs seems to result from resorption or oviposition. However, no fused eggs are observed in the ovaries, suggesting that few eggs are consumed for somatic maintenance. In such cases in which there is no addition of immature eggs during the adult stage, the decreasing number of immature eggs in the ovaries is thought to be the result of oviposition. Thus, the monogamous female of age BBB may have laid 150 eggs. Konagaya et al. (2015) also examined the number of eggs loaded in field-caught females of the pipevine swallowtail, *Battus philenor*, which is also a monandrous species, and estimated that the potential fecundity is 430 and the lifetime reproductive output is 240, although both estimates are positively related to female body size (Fig. 3.13).

Table 3.1 Number of eggs loaded in the swallowtail butterflies, *Papilio xuthus* and *Papilo machaon*, in relationship to the five age class (see p. 24) for spring and summer generations (±SE)

Spring generation						
	Age class	FF	F	B	BB	BBB
P. xuthus	No. females examined	17	19	15	8	5
	No. mature eggs	17.8 ± 4.9	25.6 ± 4.8	23.9 ± 3.5	17.8 ± 3.3	16.0 ± 9.3
	No. submature eggs	54.9 ± 6.2	55.5 ± 6.8	31.4 ± 7.6	14.1 ± 1.8	8.6 ± 1.9
	No. immature eggs	525.6 ± 49.7	443.8 ± 41.2	280.3 ± 38.2	186.0 ± 41.6	128.0 ± 19.4
	No. total eggs	602.9 ± 51.0	524.8 ± 40.7	335.6 ± 44.8	217.9 ± 41.6	152.6 ± 19.9
P. machaon	No. females examined	6	8	5	1	2
	No. mature eggs	43.5 ± 11.9	47.4 ± 6.8	43.2 ± 6.6	26	13.5 ± 6.5
	No. submature eggs	38.0 ± 5.9	22.4 ± 6.4	32.6 ± 6.0	19	8.5 ± 2.5
	No. immature eggs	394.7 ± 31.5	407.5 ± 51.0	428.8 ± 110.0	136	244.0 ± 148.0
	No. total eggs	476.2 ± 19.7	477.3 ± 52.2	504.6 ± 119.2	181	266.0 ± 157.0
Summer generation						
	Age class	FF	F	B	BB	BBB
P. xuthus	No. females examined	37	64	50	30	24
	No. mature eggs	46.8 ± 4.6	58.3 ± 4.0	47.4 ± 3.4	36.9 ± 3.2	24.7 ± 3.7
	No. submature eggs	48.4 ± 3.3	48.0 ± 3.6	38.9 ± 3.0	28.0 ± 2.4	17.2 ± 1.6
	No. immature eggs	475.0 ± 21.4	460.2 ± 19.7	367.7 ± 21.8	359.8 ± 35.2	274.5 ± 26.4
	No. total eggs	569.7 ± 21.4	567.2 ± 17.2	453.9 ± 21.8	426.0 ± 35.9	316.5 ± 28.9
P. machaon	No. females examined	15	18	6	3	2
	No. mature eggs	75.9 ± 9.2	88.7 ± 6.6	40.2 ± 9.9	66.0 ± 28.5	72.5 ± 23.5
	No. submature eggs	48.7 ± 3.8	57.3 ± 8.5	36.2 ± 3.9	42.3 ± 12.3	48.0 ± 2.0
	No. immature eggs	500.8 ± 31.2	480.1 ± 32.5	362.0 ± 43.3	391.7 ± 130.1	538.0 ± 234.0
	No. total eggs	626.2 ± 30.7	626.1 ± 32.6	438.3 ± 42.2	500.0 ± 140.4	658.5 ± 208.5

Source: Watanabe and Nozato (1986)

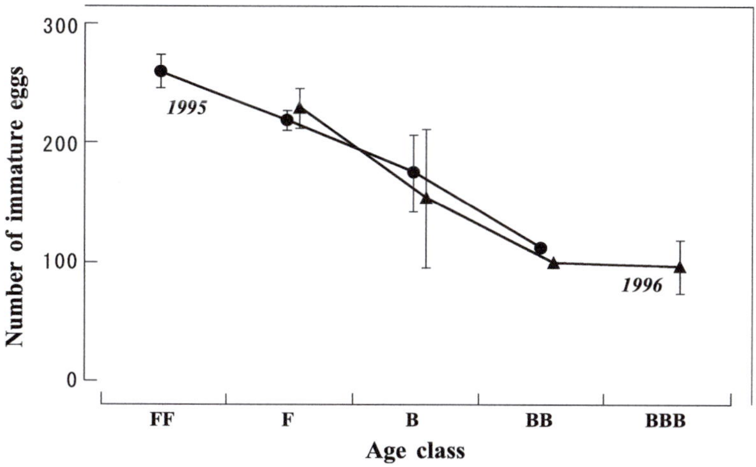

Fig. 3.12 Changes in the number of immature eggs of age FF, F, B, BB and BBB wild monandrous females of the small copper, *Lycaena phlaeas*, in 1995 (*circle*) and 1996 (*triangle*). Each *bar* represents the standard error (SE). (From Watanabe and Nishimura 2001)

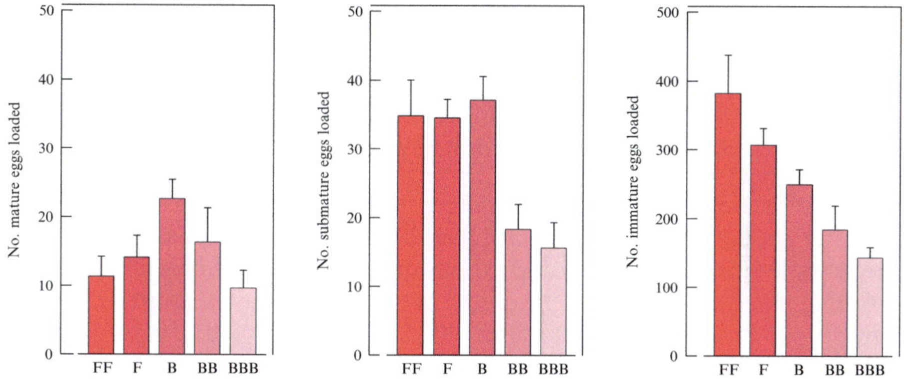

Fig. 3.13 Changes in the number of mature, submature and immature eggs with female age as indicated by wing wear in the pipevine swallowtail, *Battus philenor*. The number of mature eggs does not differ with female age. Significant changes in the number of eggs with female age are found for submature and immature eggs. The potential fecundity of females is estimated to be approximately 430 eggs: 11 mature eggs + 35 submature eggs + 383 immature eggs in FF females. (From Konagaya et al. 2015)

Although Norris (1936) found a relationship between fecundity and the quality of the nectar taken by female adults of some lepidopteran species, both ovarian capacity and daily egg production are determined by female size, ultimately derived from the quality of larval nutrition. In addition, Watanabe (1992) clarified the relationship between accumulated sugar intake and the number of mature eggs developed in older adults of the swallowtail butterfly, *Papilio xuthus*. However, the number of eggs actually laid is dependent upon mating frequency in the cabbage white, *Pieris rapae* (Watanabe and Ando 1994). Because oviposition is stimulated by the spermatophore in polyandrous species (Watanabe 1988), increases in the reproductive output resulting from multiple mating might be caused by the receipt of more oviposition stimulants. Spermatophores have been shown to increase female reproductive output in some butterfly species (Boggs and Gilbert 1979). Therefore, females in polyandrous species without early mating might lose the potential number of eggs deposited until their first mating. They seem to maintain themselves not only by using their fat bodies but also by resorption of unfertilised mature eggs, as pointed out by Dunlap-Pianka et al. (1977). Considering the stress of insemination with the sperm or the nutrients, the number of mature eggs of females which have mated sometimes will be higher than that of those females that remain unmated. The strategy of females mating as soon as possible is generally advantageous.

Eupyrene Sperm and Apyrene Sperm

Lepidoptera, including butterflies, have two distinct sperm types: one is fertile ordinary eupyrene sperm and the other is nonfertile apyrene sperm, which lacks nuclear materials. The shape and the size of eupyrene and apyrene spermatozoa of several butterfly species were reported for the first time by Tsukaguchi and Kurotsu (1922); for example, the length of an apyrene spermatozoon is 500 μm, whereas that of the eupyrene sperm bundle is 1130 μm in the Great Mormon, *Papilio memnon*. Thereafter, the polymorphic sperm types of many lepidopteran insects have been confirmed. Mancini and Dolder (2001) stated that sperm length varies considerably, from 100 μm to more than 1 mm, and that the length may be important in sperm competition because of their motility is sufficient to allow more efficiency in fertilising eggs. Accordingly, the apyrene spermatozoon is shorter than the eupyrene spermatozoon; for example, the former is 300 μm and the latter is 550 μm in the Mexican fritillary, *Euptoieta hegesia*. In the swallowtail butterflies, the length of an apyrene and an eupyrene spermatozoon is 630 μm and 917 μm, respectively, for *Papilio xuthus* (a polyandrous species), whereas lengths are 275 μm and 563 μm, respectively, for *Papilio machaon* (a monandrous species), as reported by Watanabe and Kobayashi (2006). Gage (2012) pointed out that if longer sperm allow faster swimming velocities, more powerful propulsive forces, or greater resource provision to these behaviours, then postcopulatory sexual selection could favour sperm elongation, but only so long as the female has evolved a competitive arena demanding these functions.

The flagellum of both type of sperm consists of a 9 + 9 + 2 axoneme and two mitochondrial derivatives (Figs. 3.14 and 3.15). Friedländer (1997) clarified that the two types of sperm differ considerably in the architecture of the head, the ultra-structure of the mitochondrial derivatives, and cell dimensions, among other features. In addition, the intratesticular eupyrene spermatozoon has a complicated surface pattern, consisting of the lacinate appendages and the reticular appendage,

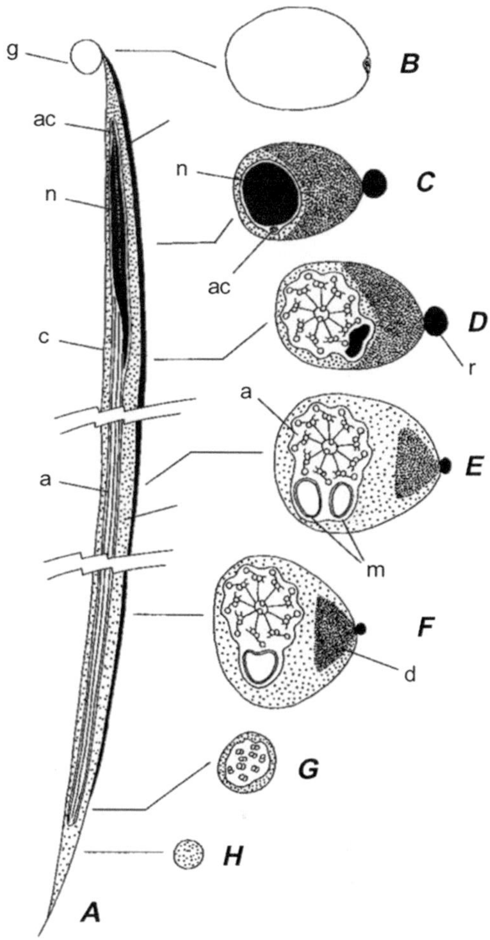

Fig. 3.14 Schematic representation of an eupyrene spermatozoon in the Mexican fritillary, *Euptoieta hegesia*: longitudinal (**a**) and transverse (**b–h**) sections. *g* globular structure, *ac* acrosome, *n* nucleus, *a* axoneme, *m* mitochondrial derivative, *r* reticular appendage, *d* dense material, *c* coat (From Mancini and Dolder 2001)

compared with the apyrene spermatozoon. The architecture of the sperm surface changes completely during passage through the male genital duct, during which the lacinate appendages surrounding the eupyrene spermatozoon are replaced.

Because the apyrene spermatozoon is, in general, shorter and thinner than eupyrene one, and also seems to contain less mitochondrial material, very early researchers regarded that apyrene spermatozoa might appear by the stimulus of an abnormal physiological environment during development or by the consequence of replication error during meiosis. However, the differentiation of eupyrene and apyrene spermatozoa in the testis, as well as their release to the vas deferens, is independently controlled, and acceleration of apyrene but not eupyrene spermatogenesis can be achieved (Friedländer 1997).

Eupyrene spermatogenesis begins during the larval phase, and ends at pupation, whereas apyrene spermatogenesis begins just before or after pupation and continues in the adult stage (Swallow and Wilkinson 2002). Then, the distribution of eupyrene

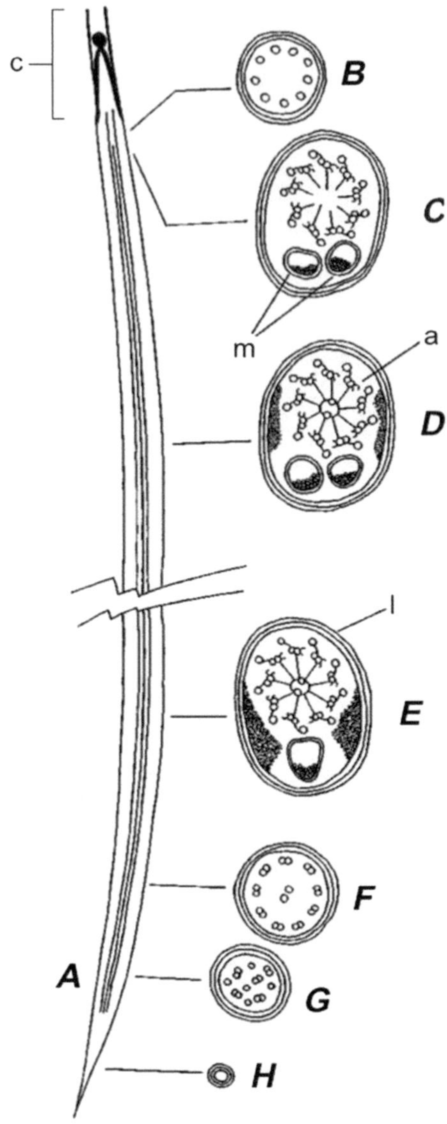

Fig. 3.15 Schematic representation of an apyrene spermatozoon in the Mexican fritillary, *Euptoieta hegesia*: longitudinal (**a**) and transverse (**b–h**) sections. *c* cap, *a* axoneme, *m* mitochondrial derivative, *l* concentric layers (From Mancini and Dolder 2001)

and apyrene sperm within the seminal vesicles differs from species to species. Both types of sperm undergo extensive metamorphosis in the testis (LaChance et al. 1977), but only eupyrene sperm continue to metamorphose after being transferred to the female. In the silk moth, *Bombyx mori*, Katsuno (1977) found that apyrene sperm bundles began to move out from the follicle at 144 h after pupation and were separated when they passed through the basement membrane, whereas the eupyrene sperm bundles separated in the copulatory pouch of the mated females. Fänger and Naumann (1993) demonstrated that apyrene sperm encircled eupyrene sperm bundles in the testis. There is a simplified mode of bipartite distribution (apyrene sperm anterior and eupyrene sperm posterior) in the males of many lepidopteran insects (Riemann et al. 1974); this asymmetrical distribution is useful in terms of the timing of transferring apyrene and eupyrene sperm bundles before the completion of copulation.

In addition, males alter the numbers of the two types of sperm ejaculated into the spermatophore independently (Wedell and Cook 1999). Furthermore, most butterfly species produce a large number of apyrene sperm, often in excess of the eupyrene sperm. Therefore, apyrene sperm must show traits behaviourally, physiologically and morphologically distinct from the eupyrene sperm (Silberglied et al. 1984). Such heteromorphic sperm have been reported for a wide variety of other invertebrates, apparently evolving independently several times in arthropods, including spiders and centipedes as well as insects (Swallow and Wilkinson 2002). Consequently, selection seems to act on sperm morphology, probably because any sperm trait that enhances competitive ability should be favoured in males possessing it.

Sperm under storage in the seminal vesicles of male are essentially motionless, and they become motile when ejaculated into the spermatophore. Shepherd (1974) stated that during transfer from male to female, the sperm become motile when they come into contact with one or more of the secretions of the male or the female accessory glands. All the factors necessary for sperm motility in the spermatophore must come from the male's reproductive tract.

During copulation, male butterflies transfer a single spermatophore, which contains both types of sperm for fertilisation as well as accessory substances. Watanabe et al. (2000) showed the ejaculation of the accessory substances as well as the number of eupyrene sperm bundles and free apyrene spermatozoa during a copulation of the swallowtail butterfly, *Papilio xuthus* (Fig. 3.16). There is no detectable weight of ejaculate in males at least 10 min after the onset of copulation. The bursa copulatrix, initially a thin, flat ellipse, grew rapidly in volume and became oval at about the time the spermatophore was structured. During the copulation, male secretions from the ductus ejaculatorius and accessory substances seem to be transferred serially to the bursa copulatrix. The ejaculate is a white gel mainly observed on the tip of the penis 20 min after the onset of copulation. No spermatophore capsule in the bursa copulatrix is structured until copulation has lasted at least 30 min. Because only a portion of these secretions is encased within a definite envelope (capsula) containing the sperm and other material, the spermatophore proper and the material lying outside the capsula are easily distinguishable. Neither eupyrene nor apyrene sperm is found in the ejaculate of the bursa copulatrix until 20 min of copulation, indicating that males did not transfer either type of sperm or accessory substances during this period. The ejaculate in cases of copulation that are interrupted after 30 min contains both apyrene and eupyrene sperm, as well as a partially structured spermatophore. Both types of sperm are observed in the spermatophore 50 min after the start of copulation, when some pairs have already completed copulation.

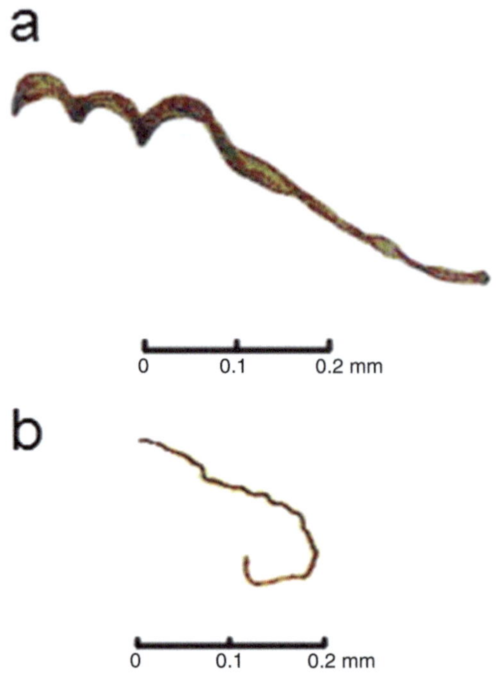

Fig. 3.16 Stereomicrograph of eupyrene sperm bundle (**a**) and apyrene spermatozoon (**b**) from a spermatophore of the swallowtail butterfly, *Papilio xuthus* (From Sasaki and Watanabe 2015)

Therefore, males may begin to ejaculate both sperm types just after the spermatophore has been structured.

In many butterfly species, there are typically 256 eupyrene spermatozoa in each bundle (Cook and Wedell 1999). The number of bundles is multiplied by 256 to give the total number of eupyrene spermatozoa. Although numerical superiority in eupyrene spermatozoa is an adaptive strategy to enhance sperm competition for the fertilisation of eggs, the males of many butterfly species transfer a larger number of free apyrene spermatozoa than that of eupyrene spermatozoa (Cook and Gage 1995). Consequently, a larger number of apyrene spermatozoa than eupyrene spermatozoa are produced within the testis; this is the case even when one multiplies the number of bundles by 256 to account for the fact that a single bundle contains 256 free eupyrene spermatozoa (Gage and Cook 1994).

Males of a species incur a substantial cost in regard to metabolic energy use in spermatophore production (Watanabe and Hirota 1999), and spermatophores persist in the reproductive tract of female butterflies. A single mating has to provide sufficient sperm in a single spermatophore to provide genetic material to all the eggs produced in a female's lifespan (Rutowski 1984). A sufficient number of eupyrene spermatozoa might migrate from the spermatophore to the spermatheca after copulation to fertilise the eggs when the female starts to oviposit.

For the swallowtail butterfly, *Papilio xuthus*, the number of eupyrene sperm bundles (Fig. 3.17) in the spermatophore just after copulation is 42 (Watanabe

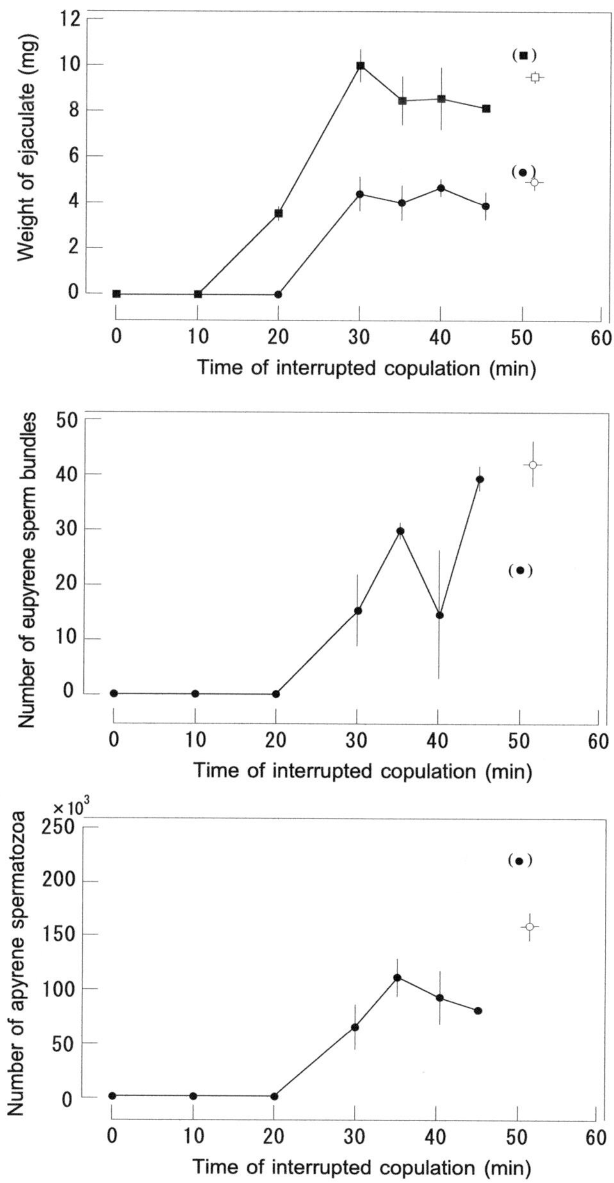

Fig. 3.17 Changes in weight of ejaculate, number of eupyrene sperm bundles, and estimated number of free apyrene spermatozoa transferred from male swallowtail butterfly, *Papilio xuthus*, after initiation of copulation (±SE). *Squares* represent the total weight of ejaculates transferred from males (accessory substances and a spermatophore); *circles* indicate weight of spermatophores. Eupyrene sperm bundles and free apyrene sperm were found in the spermatophore and not among the accessory substances. Each *open circle* and *open square* show the weight of each ejaculate in cases of uninterrupted copulation. Each set of parentheses shows one sample (From Watanabe et al. 2000)

et al. 2000). Thus, the number of bundles multiplied by the number of eupyrene spermatozoa in each bundle (256) gives a total of about 10,000 eupyrene spermatozoa that are transferred during the mating. On the other hand, a spermatophore contains about 250,000 apyrene spermatozoa just after the termination of copulation. Therefore, apyrene spermatozoa outnumber eupyrene spermatozoa in the spermatophore. During copulation, virgin males transfer apyrene spermatozoa at a rate about 25 fold that of eupyrene spermatozoa. In contrast, males of the monandrous swallowtail butterfly, *P. machaon*, transfer approximately 30,000 (= 120 bundles $\times$ 256) eupyrene spermatozoa, which can be the level guaranteeing adequate fertilisation, and 200,000 apyrene spermatozoa are transferred in the same spermatophore (Watanabe and Kobayashi 2006). Sahara and Takemura (2003) pointed out that the ratio of apyrene spermatozoa to eupyrene spermatozoa in a testis of the silkworm, *Bombyx mori*, is approximately 7:3, and that such ratio seemed to be important for successful fertilisation. The ratio of apyrene spermatozoa to eupyrene spermatozoa in a single spermatophore is 5:1 for the oriental leafworm moth, *Spodoptera litura* (Etman and Hooper 1979), 9.6:1 for the tobacco hornworm, *Manduca sexta* (Silberglied et al. 1984) and 9:1 for the Indian mealmoth, *Plodia interpunctella* (Gage and Cook 1994).

After sperm migration, in the swallowtail butterfly, *Papilio xuthus*, 1,500 eupyrene and 10,000 apyrene spermatozoa have successfully arrived in the spermatheca. Although the bursa copulatrix has a capability of elongating with the number and the volume of spermatophores containing, the spermatheca seems to remain constant in size during the female lifespan despite the number of spermatozoa contained (Watanabe et al. 2000). If the spermatheca has a capacity of a certain number of spermatozoa, excess sperm migration from the spermatophores after multiple matings may cause the spermatheca to overflow.

Again, the length of the spermatheca and of the spermathecal gland of the swallowtail butterfly, *Papilio xuthus*, is 2.8 mm and 31.0 mm, respectively (Fig. 3.6). Watanabe et al. (2000) reported that the length of both the seminal duct and the spermathecal duct in females between the bursa copulatrix and the spermatheca is about 5.2 mm. Consequently, either eupyrene or apyrene spermatozoa that had arrived at the end of the spermathecal gland have to travel at maximum 39 mm. Tschudi-Rein and Benz (1990) showed that the female pumps the sperm, so the sperm are not expending energy in transit. Watanabe and Sato (1993) suggested that sugars contained in the spermatophore might be exploited as an energy source by sperm for sperm migration and not constitute a male investment to a female. However, many spermatozoa disappear during the trip, as well.

Although there is little information on the mating frequency of male swallowtail butterflies in nature, Watanabe and Kobayashi (2006) found that males of both *Papilio xuthus* and *Papilio machaon* mate at intervals of more than a few days, extrapolated in part by their lifespan of 3–5 weeks in the laboratory. *Papilio xuthus* males produced a smaller number of large eupyrene spermatozoa, whereas *P. machaon* produced a large number of small eupyrene spermatozoa. However, there are more larger apyrene spermatozoa in *P. xuthus* than in *P. machaon*. Consequently, *P. xuthus* males must provide for the extra cost to produce eupyrene spermatozoa and ignore the cost to produce apyrene spermatozoa for each mating under severe sperm competition. By comparison, *P. machaon* males decrease the apyrene sperm mass under low sperm competition in the reproductive organs of females. Therefore, female reproductive tract architecture is one of the important determinants of the outcome of sperm competition.

Sperm Migration

After the completion of copulation, the number of eupyrene sperm bundles in a spermatophore gradually decreased. As shown in Fig. 3.18, for the swallowtail butterfly, *Papilio xuthus*, 1 h after the termination of copulation, about 17 of 41 bundles disappear; that is, if bundles are multiplied by 256, about 4,400 free eupyrene spermatozoa have been released into the spermatophore (Watanabe et al. 2000). Fifteen bundles disappear during the subsequent hour; thus, a total of approximately 8,200 eupyrene spermatozoa are in the spermatophore. Few eupyrene sperm bundles are found in the spermatophore 2 days after the completion of copulation. Thus, more than 11,000 free eupyrene spermatozoa must exist in the spermatophore within a day after copulation.

One hour after the completion of copulation, about 3200 free eupyrene spermatozoa are in the spermatophore, which roughly coincides with the calculated number of eupyrene spermatozoa from the dissolved bundles (Fig. 3.19). Approximately 8600 free eupyrene spermatozoa are in the spermatophore 2 h after completion of copulation; the number is also comparable to the number lost from eupyrene bundles 2 h after copulation, indicating that the eupyrene spermatozoa have accumulated and remain in the spermatophore. However, this number begins to decrease after 3 h, suggesting that free eupyrene spermatozoa must start to move after remaining at least 2 h in the spermatophore. Because very few free eupyrene spermatozoa are found 4 days after the copulation, all the eupyrene spermatozoa delivered from the loosened bundles must have migrated to the spermatheca.

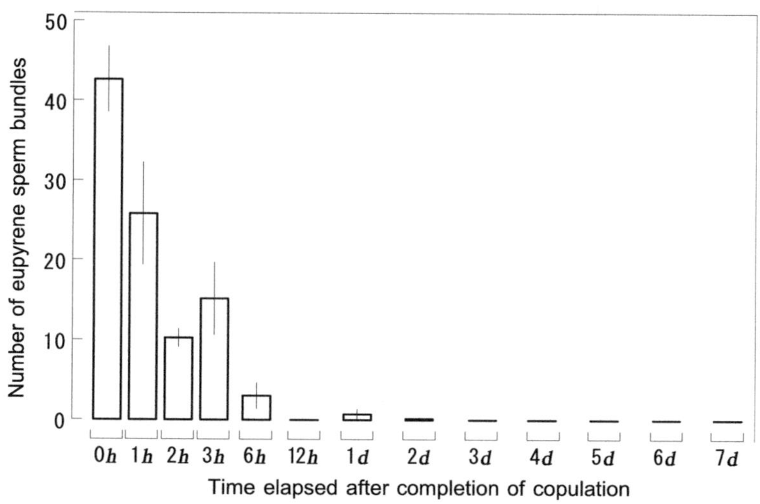

Fig. 3.18 Change in the number of eupyrene sperm bundles in the spermatophore for the female swallowtail butterfly, *Papilio xuthus*, after completion of copulation (±SE). *0 h* shows the number of eupyrene sperm bundles immediately after completion of copulation, indicating that males transferred 42.7 bundles (From Watanabe et al. 2000)

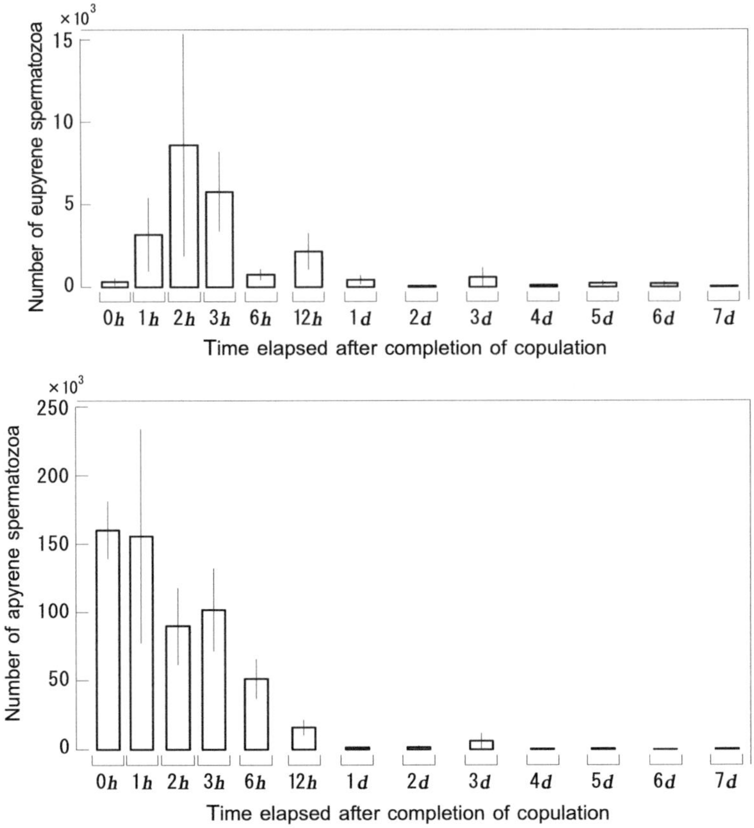

Fig. 3.19 Changes in estimated number of free eupyrene and free apyrene spermatozoa in the spermatophore for the female swallowtail butterfly, *Papilio xuthus*, after completion of copulation (±SE). *0 h* shows the number of spermatozoa immediately after completion of copulation, indicating that males transferred about 160,000 free apyrene spermatozoa (Watanabe et al. 2000)

As time passes, the number of apyrene spermatozoa in the spermatophore also decrease, though they remain about 1 h in the spermatophore after the completion of copulation. Generally, the spermatophore lacks apyrene spermatozoa 12 h after copulation. Because females can begin to absorb ejaculate 3 days after copulation (Watanabe et al. 2000), this decline is the result of the migration of apyrene spermatozoa from spermatophore to spermatheca.

The sperm moving out from the spermatophore inside the bursa copulatrix migrate as a mass, as observed in the silkworm, *Bombyx mori* (Suzuki et al. 1996) and in the swallowtail butterfly, *Papilio xuthus* (Watanabe and Sasaki 2010). The mass of sperm first enters the duct seminalis, and thereafter moves into the spermathecal duct across the vestibulum. When the female is mated, the mass of sperm that had previously entered the spermatheca is gradually forced to the inner part of the spermatheca, probably by the apyrene sperm, because the apyrene sperm

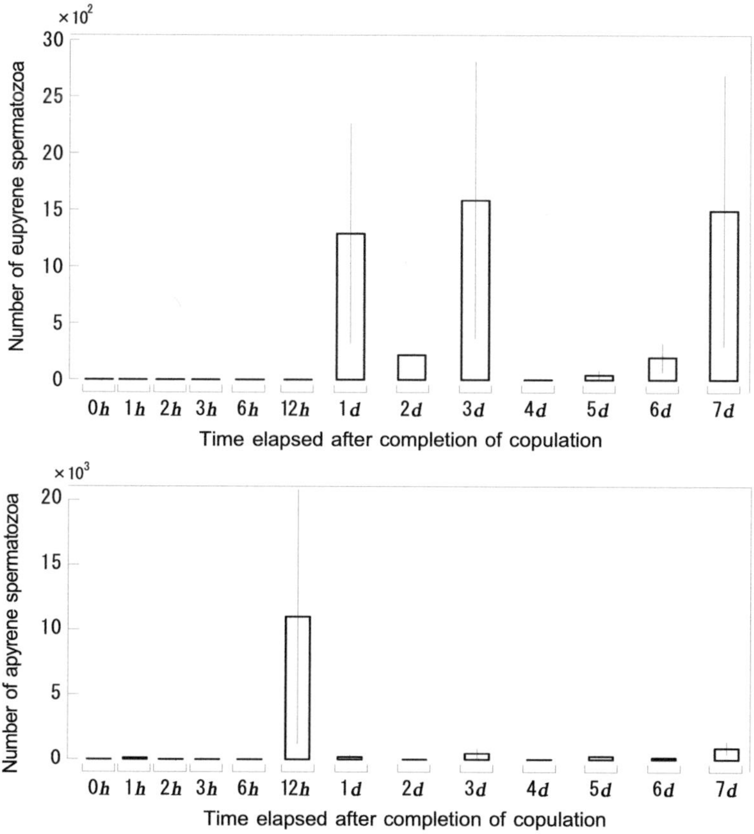

Fig. 3.20 Changes in estimated number of free eupyrene and free apyrene spermatozoa in the spermatheca for the female swallowtail butterfly, *Papilio xuthus*, after the completion of copulation (±SE) (From Watanabe et al. 2000)

migration takes place earlier than that of the eupyrene sperm. Consequently, the eupyrene sperm that enter later would occupy space near the entrance of the spermatheca, indicating that the eupyrene sperm could easily fertilise eggs.

Although the bursa copulatrix of females is balloon shaped and can expand to several times its original size from the number of spermatophores, the spermatheca body does not change in size, even when the female is polyandrous.

Rapid arrival of the sperm at the fertilisation point appears to be a ubiquitous phenomenon among many insect species (Sahara and Takemura 2003). As shown in Fig. 3.20, in the spermatheca of the swallowtail butterfly, *Papilio xuthus*, a few apyrene spermatozoa arrive 6 h after copulation ceased, and after 12 h, more than 11,000 apyrene spermatozoa are counted (Watanabe et al. 2000). However, the number of apyrene spermatozoa in the spermatheca decreases 1 day after the copulation, and apyrene spermatozoa disappear from the spermatheca. Therefore, after 2 days, the apyrene spermatozoa could be found in neither the spermatophore nor the spermatheca. More than 90 % of the apyrene spermatozoa disappeared during the sperm migration, probably because of absorption by the female. On the other hand, after 1 day about 1,300 eupyrene spermatozoa are found in the spermatheca. Not all the eupyrene spermatozoa also have arrived in the spermatheca 1 day after the completion of copulation. Therefore, it is assumed that apyrene and eupyrene spermatozoa should reach the spermatheca from the spermatophore 12 h and more than 1 day, respectively, after copulation ceases. About 1,000 eupyrene spermatozoa are able to stay alive for at least 7 days in the spermatheca of females.

Sperm migration from the bursa copulatrix to the spermatheca is affected by females, probably by the muscular control in Lepidoptera (LaMunyon and Eisner 1993). Watanabe and Hachisuka (2005) showed that only 10 % of eupyrene spermatozoa and 0.4 % of apyrene spermatozoa had arrived in the spermatheca. In the cotton bollworm, *Helicoverpa armigera*, females always stored 30,000 eupyrene spermatozoa and 160,000 apyrene spermatozoa, irrespective of the mating number (Yan et al. 2013). The difference in sperm number between the spermatophore and the spermatheca could result from use as egg nutrients (Riemann et al. 1974) or for female body maintenance (Silberglied et al. 1984), and it could also be the result of their own mortality (Yamagishi et al. 1992). Labine (1966) stated that even if as few as about 10 % of the eupyrene spermatozoa is transferred by the male, the number would still be sufficient to fertilise all the eggs loaded in the female. Tschudi-Rein and Benz (1990) pointed out that in butterflies the female appears to control the sperm migration process to the spermatheca.

The physiological and morphological mechanisms of apyrene sperm loss from the spermatheca have not yet been investigated. However, this loss may be caused by the filling of the vacant top position of the cul-de-sac filament of the spermathecal gland by apyrene sperm. It is also possible that apyrene sperm are absorbed by the female. The long filament associated with the tip of the spermatheca may affect the distribution of sperm in the ductus seminalis; this would describe, for example, the function of the spermatheca of the swallowtail butterfly, *Papilio xuthus*. On the other hand, Etman and Hooper (1979) suggested that spermathecal muscles may regulate the movement of sperm from the spermatheca.

Because apyrene spermatozoa arrive in the spermatheca earlier than eupyrene spermatozoa, the spermathecal gland is first filled by apyrene spermatozoa. However, females being allowed to oviposit have a lower number of apyrene spermatozoa in the gland than those inhibited from oviposition, suggesting that apyrene spermatozoa might be used in oviposition activity. On the other hand, few eupyrene spermatozoa are present in the spermathecal gland of females being inhibited from oviposition, probably because of the large number of apyrene spermatozoa therein.

Storage of a sufficient number of sperm in the spermatheca is important for females to fertilise eggs. When sperm depletion has occurred, females cannot oviposit fertilised eggs. Because polygamous females could regain eupyrene sperm repeatedly by multiple mating, the risk of sperm depletion might be decreased. In contrast, for monogamous females, the cost of re-mating such as the waste of time or energy, and the risk of predation, disease and harassments by the other males, must be greater than the risk of sperm depletion, as reported by

Arnqvist and Rowe (2005). However, Elzinga et al. (2011) mentioned that sperm depletion in monogamous females is critical for their reproductive success, and that they should store a larger number of sperm than the fecundity available at mating.

References

Ando S, Watanabe M (1992) Egg load and multiple matings of a cabbage butterfly, *Pieris canidia indica* (Evans), in the wild. Jpn J Appl Entomol Zool 36:200–201. (in Japanese with English summary)

Ando S, Watanabe M (1993) Mating frequency and egg load in the white butterfly, *Pieris melete* Menetries, in a wild environment. Jpn J Ecol 43:111–114 (in Japanese)

Arnqvist G, Rowe L (2005) Sexual conflict. Princeton University Press, Princeton and Oxford

Boggs CL, Gilbert LE (1979) Male contribution to egg production in butterflies: evidence for transfer of nutrients at mating. Science 206:83–84

Cook PA, Gage MJG (1995) Effects of risks of sperm competition on the numbers of eupyrene and apyrene sperm ejaculated by moth *Plodia interpunctella* (Lepidoptera: Pyralidae). Behav Ecol Sociobiol 36:261–268

Cook PA, Wedell N (1999) Non-fertile sperm delay female remating. Nature 397:486

Drummond BA III (1984) Multiple mating and sperm competition in the Lepidoptera. In: Smith RL (ed) Sperm competition and the evolution of animal mating systems. Academic Press, New York, pp 291–370

Dunlap-Pianka H, Boggs CL, Gilbert LE (1977) Ovarian dynamics in heliconiine butterflies: programmed senescence versus eternal youth. Science 197:487–490

Elzinga JA, Chevasco V, Grapputo A, Mappes J (2011) Influence of male mating history on female reproductive success among monandrous Naryciinae (Lepidoptera: Psychidae). Ecol Entomol 36:170–180

Etman AAM, Hooper GHS (1979) Sperm precedence of the last mating in *Spodoptera litura*. Ann Entomol Soc Am 72:119–120

Fänger H, Naumann CM (1993) Correlation between the mesodermal male genital ducts and the spermatophore structure in a ditrysian moth, *Zygaena trifolii* (Esper, 1783) (Insecta, Lepidoptera, Zygaenidae). Acta Zool 74:239–246

Feltwell J (1986) The natural history of butterflies. Croom Helm, London

Friedländer M (1997) Control of the eupyrene–apyrene sperm dimorphism in Lepidoptera. J Insect Physiol 43:1085–1092

Gage MJG (2012) Complex sperm evolution. Proc Natl Acad Sci USA 109:4341–4342

Gage MJG, Cook PA (1994) Sperm size or numbers? Effects of nutritional stress upon eupyrene and apyrene sperm production strategies in the moth *Plodia interpunctella* (Lepidoptera: Pyralidae). Funct Ecol 8:594–599

Galicia I, Sánchez V, Cordero C (2008) On the function of signa, a genital trait of female Lepidoptera. Ann Entomol Soc Am 101:786–793

He Y, Tanaka T, Miyata T (1995) Eupyrene and apyrene sperm and their numerical fluctuations inside the female reproductive tract of the armyworm, *Pseudaletia separata*. J Insect Physiol 41:689–694

Hiroyoshi S, Mitsuhashi J (1999) Sperm reflux and its role in multiple mating in males of a butterfly *Polygonia c-aureum Linnaeus* (Lepidoptera: Nymphalidae). J Insect Physiol 45:107–112

Katsuno S (1977) Studies on eupyrene and apyrene spermatozoa in the silk worm, *Bombyx mori* L. (Lepidoptera: Bombycidae). II. The intratesticular behaviour of the spermatozoa after emergence. Appl Entomol Zool 12:236–240

Konagaya T, Mutoh N, Suzuki M, Rutowski RL, Watanabe M (2015) Estimates of female lifetime fecundity and changes in the number and types of sperm stored with age and time since mating in the monandrous swallowtail butterfly, *Battus philenor* (Lepidoptera: Papilionidae) in the Arizona desert. Appl Entomol Zool 50:311–316

Labine PA (1966) The population biology of the butterfly, *Euphydryas editha*. IV. Sperm precedence: a preliminary report. Evolution 20:580–586

LaChance LE, Richard RD, Ruud RL (1977) Movement of eupyrene sperm bundles from the testis and storage in the ductus ejaculatoris duplex of the male pink bollworm: effects of age, strain, irradiation, and light. Ann Entomol Soc Am 70:647–651

LaMunyon CW, Eisner T (1993) Postcopulatory sexual selection in an arctiid moth (*Utetheisa ornatrix*). Proc Natl Acad Sci USA 90:4689–4692

Lewis Z, Wedell N (2009) Male moths reduce sperm investment in relatives. Anim Behav 77:1547–1550

Lincango P, Fernández G, Baixeras J (2013) Microstructure and diversity of the bursa copulatrix wall in Tortricidae (Lepidoptera). Arthropod Struct Dev 42:247–256

Mancini K, Dolder H (2001) Ultrastructure of apyrene and eupyrene spermatozoa from the seminal vesicle of *Euptoieta hegesia* (Lepidoptera: Nymphalidae). Tissue Cell 33:301–308

Nabi MN, Harrison RA (1983) Activity of sperm and fertility in the potato moth, *Phthorimaea operculella*. J Insect Physiol 29:431–435

Nakanishi Y, Watanabe M, Ito T (1996) Differences in lifetime reproductive output and mating frequency of two female morphs of the sulfur butterfly, *Colias erate* (Lepidoptera: Pieridae). J Res Lepid 35:1–8

Norris MJ (1936) The feeding-habits of the adult Lepidoptera Heteroneura. Trans R Entomol Soc Lond 85:61–90

Porter K (1992) Eggs and egg-laying. In: Dennis RLH (ed) The ecology of butterflies in Britain. Oxford University Press, Oxford, pp 46–72

Riemann JG, Thorson BJ, Ruud RL (1974) Daily cycle of release of sperm from the testes of the Mediterranean flour moth. J Insect Physiol 20:195–207

Rutowski RL (1984) Sexual selection and the evolution of butterfly mating behavior. J Res Lepid 23:125–142

Sahara K, Takemura Y (2003) Application of artificial insemination technique to eupyrene and/or apyrene sperm in *Bombyx mori*. J Exp Zool A 297:196–200

Sánchez V, Hernández-Baños BE, Cordero C (2011) The evolution of a female genital trait widely distributed in the Lepidoptera: comparative evidence for an effect of sexual coevolution. PLoS ONE 6:e22642

Sasaki N, Watanabe M (2015) Alternative ejaculate allocation tactics in relation to male mating history of the swallowtail butterfly, *Papilio xuthus* L. (Lepidoptera: Papilionidae). J Res Lepid 48:1–8

Sasaki N, Konagaya T, Watanabe M, Rutowski RL (2015) Indicators of recent mating success in the pipevine swallowtail butterfly (*Battus philenor*) and their relationship to male phenotype. J Insect Physiol 83:30–36

Shepherd JG (1974) Activation of saturniid moth sperm by a secretion of the male reproductive tract. J Insect Physiol 20:2107–2122

Silberglied RE, Shepherd JG, Dickinson JL (1984) Eunuchs: the role of apyrene sperm in Lepidoptera? Am Nat 123:255–265

Sugawara T (1979) Stretch reception in the bursa copulatrix of the butterfly, *Pieris rapae crucivora*, and its role in behaviour. J Comp Physiol 130:191–199

Suzuki N, Okuda T, Shinbo H (1996) Sperm precedence and sperm movement under different copulation intervals in the silkworm, *Bombyx mori*. J Insect Physiol 42:199–204

Swallow JG, Wilkinson GS (2002) The long and short of sperm polymorphisms in insects. Biol Rev 77:153–182

Tschudi-Rein K, Benz G (1990) Mechanisms of sperm transfer in female *Pieris brassicae* (Lepidoptera: Pieridae). Ann Entomol Soc Am 83:1158–1164

Tsukaguchi R, Kurotsu T (1922) On the apyrene spermatozoa of the Lepidoptera. Zool Mag 34:310–317 (in Japanese)

Watanabe M (1981) Population dynamics of the swallowtail butterfly, *Papilio xuthus* L., in a deforested area. Res Popul Ecol 23:74–93

Watanabe M (1988) Multiple matings increase the fecundity of the yellow swallowtail butterfly, *Papilio xuthus* L., in summer generations. J Insect Behav 1:17–29

Watanabe M (1992) Egg maturation in laboratory-reared females of the swallowtail butterfly, *Papilio xuthus* L. (Lepidoptera: Papilionidae), feeding on different concentration solutions of sugar. Zool Sci 9:133–141

Watanabe M, Ando S (1993) Influence of mating frequency of lifetime fecundity in wild females of the small white *Pieris rapae* (Lepidoptera, Pieridae). Jpn J Entomol 61:691–696

Watanabe M, Ando S (1994) Egg load in wild females of the small white *Pieris rapae crucivora* (Lepidoptera, Pieridae) in relation to mating frequency. Jpn J Entomol 62:293–297

Watanabe M, Hachisuka A (2005) Dynamics of eupyrene and apyrene sperm storage in ovipositing females of the swallowtail butterfly *Papilio xuthus* (Lepidoptera: Papilionidae). Entomol Sci 8:65–71

Watanabe M, Hirota M (1999) Effects of sucrose intake on spermatophore mass produced by male swallowtail butterfly *Papilio xuthus* L. Zool Sci 16:55–61

Watanabe M, Kobayashi T (2006) Total sperm ejaculation in monandrous (*Papilio machaon*) and polyandrous (*P. xuthus*) swallowtail butterflies (Lepidoptera: Papilionidae) restricted to larval stage-derived nutrients. J Res Lepid 39:1–7

Watanabe M, Nishimura M (2001) Reproductive output and egg maturation in relation to mate-avoidance in monandrous females of the small copper, *Lycaena phlaeas* (Lycaenidae). J Lepid Soc 54:83–87

Watanabe M, Nozato K (1986) Fecundity of the yellow swallowtail butterflies, *Papilio xuthus* and *P. machaon hippocrates*, in a wild environment. Zool Sci 3:509–516

Watanabe M, Sasaki N (2010) Pattern of sperm storage and migration in the reproductive tract of the swallowtail butterfly *Papilio xuthus*: cryptic female choice after second mating. Physiol Entomol 35:328–333

Watanabe M, Sato K (1993) A spermatophore structured in the bursa copulatrix of the small white *Pieris rapae* (Lepidoptera, Pieridae) during copulation and its sugar content. J Res Lepid 32:26–36

Watanabe M, Nozato K, Kiritani K (1986) Studies on ecology and behavior of Japanese black swallowtail butterflies (Lepidoptera: Papilionidae). 5. Fecundity in summer generations. Appl Entomol Zool 21:448–453

Watanabe M, Wiklund C, Bon'no M (1998) The effect of repeated matings on sperm numbers in successive ejaculates of the cabbage white butterfly *Pieris rapae* (Lepidoptera: Pieridae). J Insect Behav 11:559–570

Watanabe M, Bon'no M, Hachisuka A (2000) Eupyrene sperm migrates to spermatheca after apyrene sperm in the swallowtail butterfly, *Papilio xuthus* L. (Lepidoptera: Papilionidae). J Ethol 18:91–99

Wedell N, Cook PA (1999) Strategic sperm allocation in the small white butterfly *Pieris rapae* (Lepidoptera: Pieridae). Funct Ecol 13:85–93

Wickman P-O (1985) Male determined mating duration in butterflies? J Lepid Soc 39:341–342

Yamagishi M, Ito Y, Tsubaki Y (1992) Sperm competition in the melon fly, *Bactrocera cucurbitae* (Diptera: Tephritidae): effects of sperm "longevity" on sperm precedence. J Insect Behav 5:599–608

Yan S, Li H, Zhang J, Zhu J, Zhang Q, Liu X (2013) Sperm storage and sperm competition in the *Helicoverpa armigera* (Lepidoptera: Noctuidae). J Econ Entomol 106:708–715

Chapter 4
Necessary Requirements for Oviposition

Keywords Fecundity • Kin • Mate refusal posture • Nuptial gift • Overwinter • Pollen • Population density • Radioactive label • Tannin • Wing dimorphism

A deposited female egg is tens of thousands of times larger than a male spermatozoon. The highest fecundity found immediately after emergence has suggested that no additional eggs develop throughout the female adult stage. Although the number of eggs actually laid during the female reproductive period is about 500, her abdomen size generally limits the number of mature eggs loaded to fewer than 50. Therefore, females have to continue to lay eggs with developing mature eggs during their lifespan, resulting in a relatively low daily number of eggs laid. Because adult butterflies feed on nectar that contains sugars with little protein, the nutrients used for egg development are those from the fat body accumulated in the abdomen, derived from host plants during larval stage. Because there is no intake of additional nutrients from nectar except sugars throughout the adult stage, using nutrients from the larval host plants appears to be a trade-off relationship between egg development and somatic maintenance. In laboratory experiments, a single mated female pierid butterfly laid eggs every day, and the egg size decreased with each day. However, females could obtain alternative nutrients for egg development and somatic maintenance throughout the adult stage, in which the nutrients are obtained from the ejaculates derived by males when copulating. A spermatophore is a nutrient-rich investment of the males that includes amino acids and sugars. Therefore, females tend to mate more than once to increase the number of eggs actually laid. Because copulating pairs are inactive and can be easily attacked by predators such as birds, multiple matings must increase the mortality risk of the females as well as of the males. Consequently, the optimal number of mating frequencies in females could be the result of a trade-off between the intake of nutrients and mortality risk. In fact, the number of lifetime matings in pierid butterfly females is similar among generations, years and various population densities.

M. Watanabe, *Sperm Competition in Butterflies*, Ecological Research Monographs,
DOI 10.1007/978-4-431-55945-0_4

Oviposition

The larvae of most butterflies feed on plants, and most of them can only feed on a few particular kinds of plant species, indicating that a newly hatched caterpillar from an egg deposited on the wrong host plants would starve to death because of its low mobility. Generally, the small creeping caterpillar itself has little chance to search for the appropriate host plant, even though such plants are sufficiently available, distributed within the habitat. Watanabe and Omata (1978) measured that the newly hatched caterpillar of the blue hairstreak, *Artopoetes pryeri*, walked 13.7 cm from the egg mass on the twig of the host tree, *Ligustrum obtusifolium*, to the buds to enter (maximum, 62.0 cm), after overwintering at the egg stage (Fig. 4.1). Therefore, parental care for most butterflies consists of the females finding the right host plant. There have been many reports on oviposition behaviour of butterflies, focussing on searching flights in the landscape as well as in the structure of plant communities (Watanabe 1979), searching images to avoid plants on which larval growth and development is poor (Rausher 1978), chemical cues (Honda and Hayashi 1995), availability of host plant quality (Watanabe 1983), female decisions in egg laying (Porter 1992) and so on (Fig. 4.2). The large white, *Pieris brassicae*, detects cabbage fields by following upwind the sulphur-rich odours emitted from the plants (Feltwell 1986). They flutter at a low height above the cabbages and seem to use this method of olfactory detection very successfully.

Because a certain variation in leaf quality is a defensive agent against herbivorous insects, a protracted larval period is the first sign of impaired leaf quality. Such a longer larval period resulting from impaired leaf quality per se may likely result in a higher risk of predation and parasitism by natural enemies. Unhealthy larvae of the swallowtail butterfly, *Papilio xuthus*, cannot avoid even small-sized predators, although they have a pair of orange horns with osmetria that produce a strong smell after having been disturbed (Watanabe 1981). A strong wind also blows such unhealthy larvae off the tall trees of the host tree, *Zanthoxylum ailanthoides*. Feeny (1970) reported that tannins of oak leaves tend to be concentrated in the

Fig. 4.1 An egg mass of the blue hairstreak, *Artopoetes pryeri*, on the twig of the host tree, *Ligustrum obsifolium* (Photograph courtesy of T. Kasui)

palisade cells of parenchyma, and that one function of leaf tannins may be to inhibit lepidopteran larvae from attack not only by reducing the nutritive value but also by influencing palatability. Watanabe (1982) clarified that the major function of tannins is to toughen the leaf by concentrating them in the parenchyma, serving to defend leaves against swallowtail butterfly larvae. The main mortality factor of the first-instar larvae reared in the laboratory is the tough leaves of the tall host trees from accumulated tannins (Fig. 4.3). Leaves of the short trees have few tannins. Thus, ovipositing females of swallowtail butterflies in the field usually avoid tall host plants on which larval growth and development is poor and search preferentially for suitable host plants (= young, short trees) on which growth and development is good. Consequently, spatial distribution of eggs of the swallowtail butterfly in a citrus grove is always contagious in new leaves among the citrus trees. In Japan, similarly, the larvae of several swallowtail butterfly species often injure the young branches of agricultural citrus trees, becoming a pest insect. Females of the lime swallowtail, *Papilo demoleus*, also laid eggs on the new leaves of young lemon trees, *Citrus limon*, and the larvae sometimes kill the host tree by defoliation caused by their feeding (Sharifi and Zarea 1970).

Different butterfly species might evolve physiological and behavioural adaptations in response to the distribution and abundance of their host plants as well as habitat suitability, density of competitors and natural enemies. Although daily weather conditions may affect the daily oviposition activity of females, resulting in variation in the daily number of eggs laid, females of most species can be expected to reach a maximum rate of eggs laid per day during their younger days because their mortality increases with age after eclosion (Fig. 4.4). Except for the zebra longwing, *Heliconius charitonius*, whose longevity is more than 3 months, the rate of egg laying reaches a peak a few days after emergence, with longevity being less than 1 month in most species (Fig. 4.5).

An egg-laying female needs to spend a few seconds actually placing her single egg. Their powerful flight performance allows females to disperse their eggs over a large area, probably to spread the risks of predation and reduce competition from conspecifics. For the swallowtail butterfly, *Papilio xuthus*, Watanabe (1979) reported that females search for suitable oviposition sites using flyways formed in the margins of forests mainly by the males, along which the available young, short host trees, *Zanthoxylum ailanthoides*, could grow as a pioneer tree. In addition, females may disperse rather than males and probably re-mate when arriving in a new locality (Fig. 4.6). Consequently, her subsequent eggs will be fathered by a male in the new locality, where the probability of survival in the deposited eggs might increase. Such an increase should occur even for individuals dispersing late in the flight season. Shapiro (1970) suggested that newly mated females of the checkered white, *Pontia protodice*, leave areas of high population density to avoid excessive male harassment. Comparable results have been suggested by Zalucki and Kitching (1982) for the monarch butterfly, *Danaus plexippus*. Considering the habitat preference of the swallowtail butterfly under plant succession (Watanabe 1981), this strategy seem to be adaptive for the use of ephemeral habitats, the initial stage of the secondary succession.

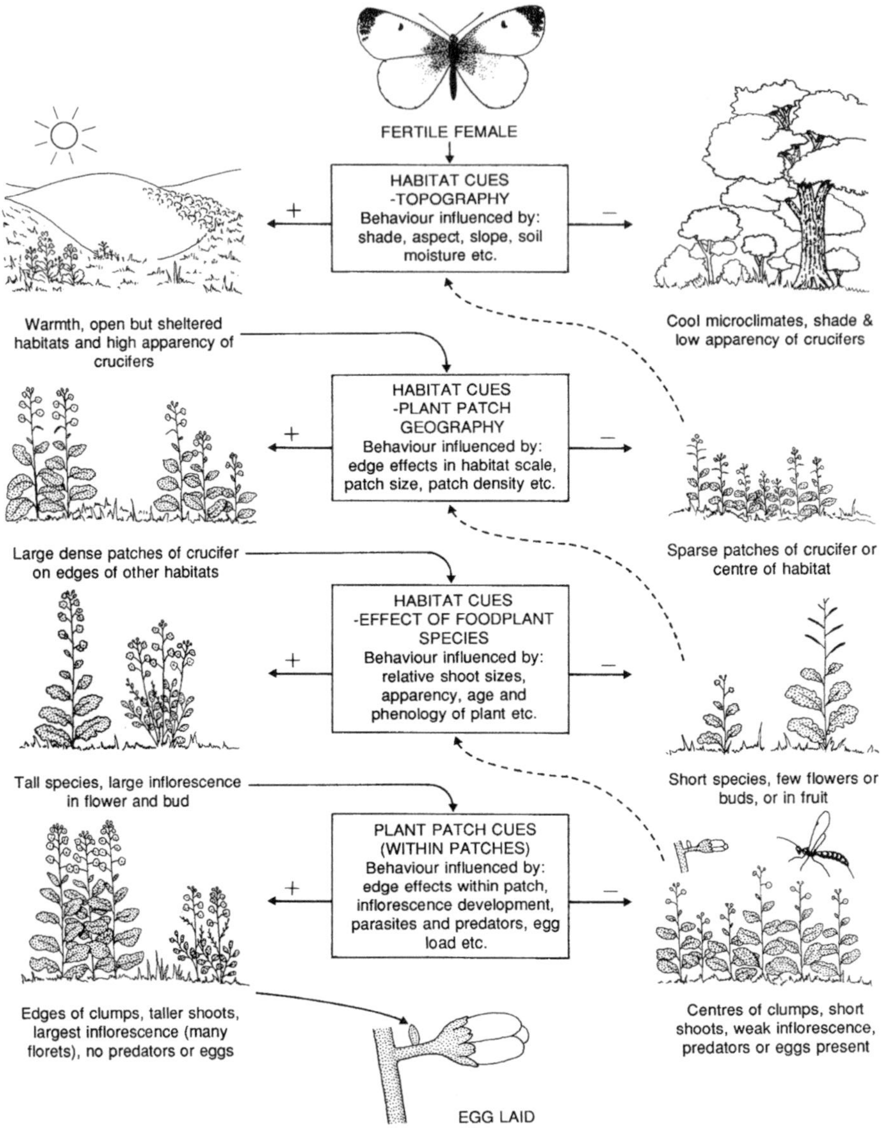

Fig. 4.2 Pathways of 'decision' for a fertile female orange-tip, *Anthocharis cardamines*. Her decisions before laying an egg can be related to many variables (given in the *central boxes*). This diagram leads through the stages of within-habitat cues, between-patch cues, cues between different host plants, and finally cues within a single host plant patch. This model provides a basic plan to describe how the butterfly may choose an egg site. The host plant species chosen and the details of selection will be influenced by the phenology of potential host plants and their relative physiological state (From Porter 1992)

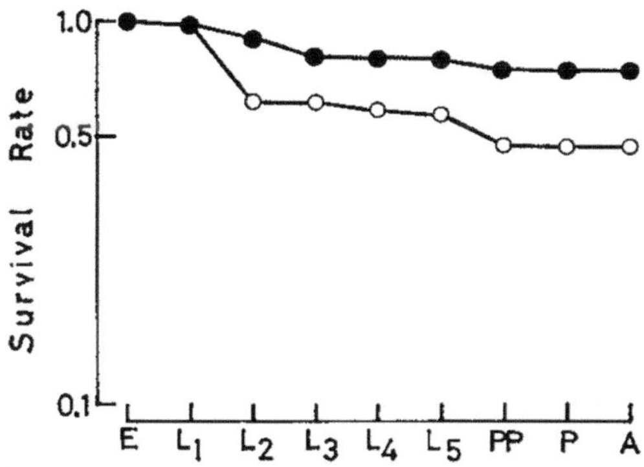

Fig. 4.3 Survivorship curves of the swallowtail butterfly, *Papilio xuthus*, reared on leaves of short (*closed circles*) and tall (*open circles*) host trees, *Zanthoxylum ailanthoides*, in the laboratory. Significant difference in the survival rate is found in the first-instar stage and in total mortality of the larval stage (From Watanabe 1982)

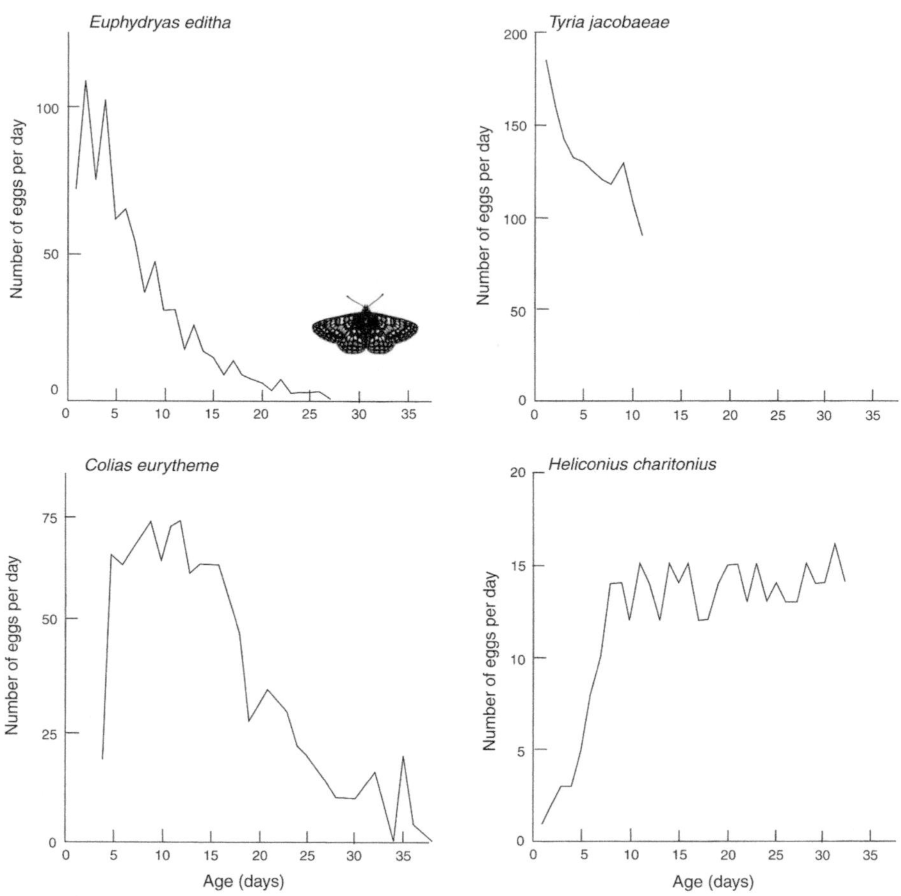

Fig. 4.4 Age-specific fecundity curves for females (From Labine 1968)

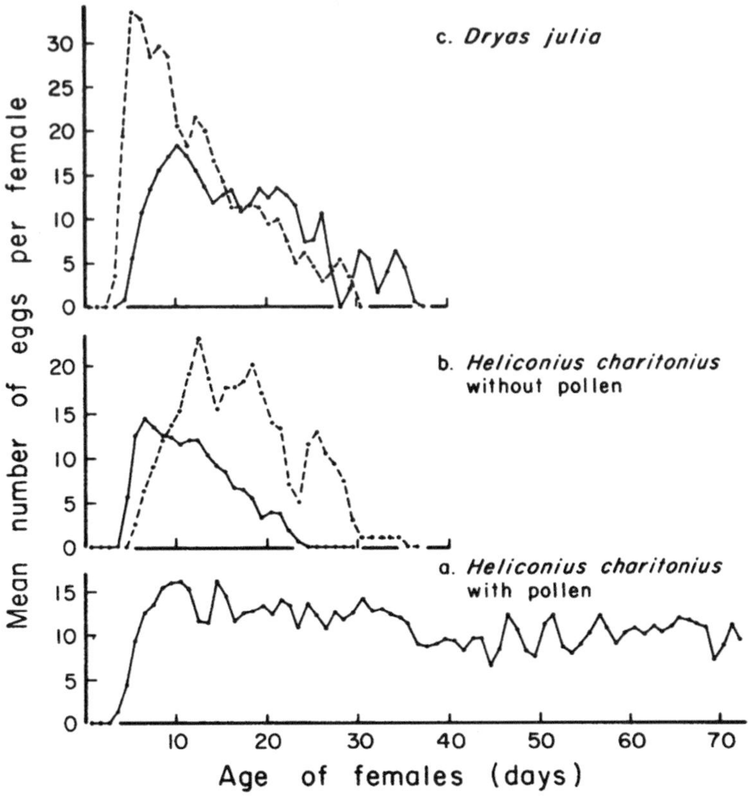

Fig. 4.5 Daily oviposition patterns in heliconiine butterflies. The *broken line* in **b** represents the means of two large females; that in **c** shows means of two small individuals. Each *solid curve* averages combined data from several females, with each point representing a 2-day moving average (From Dunlap-Pianka et al. 1977)

A female reproductive strategy includes not only her mating habits but also her age-specific oviposition schedule and oviposition habits. The age-specific oviposition schedule is the result of female allocation of nutrients from the larval stage, her adult self and male sources, to reproduction and other activities, combined with the characteristic ovarian dynamics. Oviposition pattern and longevity of pierid butterflies in the field have been described by Ohsaki (1980). Laboratory studies showed that daily oviposition of the females of the cabbage white, *Pieris rapae*, that mates only once, reaches a peak on the third day after emergence and then declines gradually (Suzuki 1978). In the green-veined white, *Pieris napi*, the female egg-laying rate peaks after 3 days since the previous mating, with polyandrous females having higher lifetime fecundity and longer lifespan than monandrous females because of the excess of male-derived accessory gland products transferred with a spermatophore at mating (Wiklund et al. 1993).

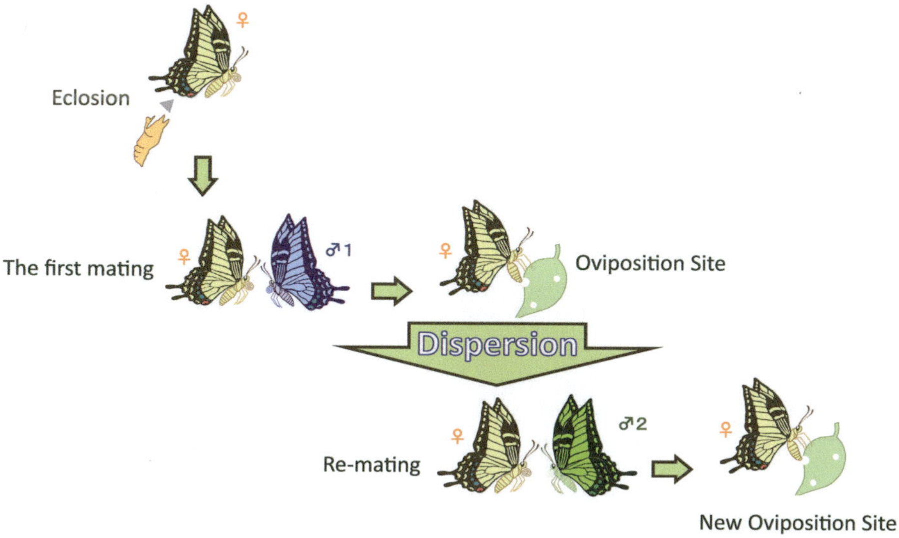

Fig. 4.6 Schematic representation of female reproductive life after eclosion in polyandrous butterflies

Variation in the numbers of mature eggs developing in the ovaries may occur as a result of a different oviposition pattern and process of egg development in the ovaries. Thus, diurnal oviposition activity in the field could be estimated by the changes in the number of mature eggs in the ovaries with time of the day. The number of mature eggs contained in the female must be high before oviposition and low after oviposition, respectively. Watanabe and Ando (1994) tried to capture the cabbage white, *Pieris rapae*, in the field throughout the day, including midnight, and clarified that when young females mate once, the number of mature eggs contained is about 75 in the morning, and decreases towards afternoon (25 eggs), indicating that females generally oviposit actively in the morning. The estimated daily number of eggs laid is 50. The number of mature eggs in the female recovers during the night (Fig. 4.7). For the swallowtail butterfly, *Papilio xuthus*, the younger females have a clear tendency for active oviposition in the morning (Watanabe and Nozato 1986). As shown in Fig. 4.8, females in the summer generation deposit 25, 25, 25, 15 and 20 eggs, in age classes FF, F, B, BB and BBB, respectively, in a day (see p. 24). However, the oviposition time of the black swallowtail butterflies, *Papilo helenus* and *Papilo protenor*, ranges from afternoon to evening (Watanabe et al. 1986).

The muscular activity of adult butterflies uses energy during flying, and this prolonged activity can be increased if the butterflies have access to enough sugars in the nectar. Nectar is mainly composed of carbohydrates, sugars that are easily converted into energy for adult activities. Watanabe et al. (1988) found three sugars, glucose, fructose, and sucrose, in the flower of the glory tree, *Clerodendron trichotomum*, a main nectar plant of swallowtail butterflies in the summer season of the warm temperate zone of Japan. Although adults of most butterfly species search

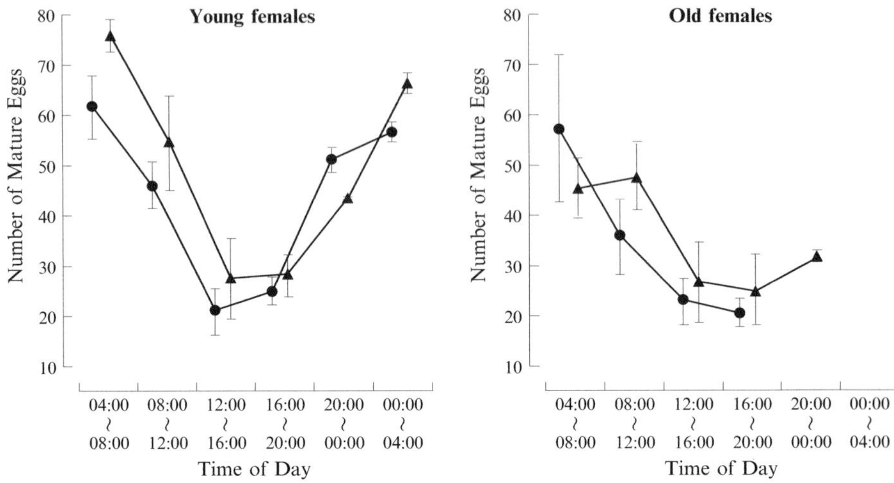

Fig. 4.7 Daily changes in the number of mature eggs loaded for monandrous (*circles*) and polyandrous (*triangles*) females of the cabbage white, *Pieris rapae*, for young and old ages (From Watanabe and Ando 1994)

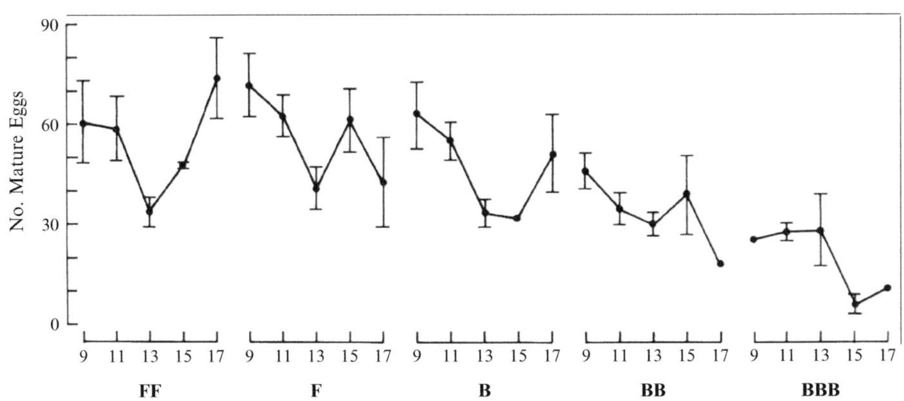

Fig. 4.8 Change in the number of mature eggs loaded in a day of the swallowtail butterfly, *Papilio xuthus*, of respective ages in the summer generations (From Watanabe and Nozato 1986)

flowers for nectar during much of the day, they do not gain the amino acids or protein which must be used for somatic maintenance and female egg development. Therefore, female reproductive output may be primarily dependent upon nitrogenous reserves accumulated during larval feeding and stored in the fat body, because no more protein-rich food for oviposition is supplied from the nectar. Females should have to acquire an alternative nutritional source for egg development in addition to the host plants during their larval stage. Therefore, female reproductive success depends on the amount of three potential resources available for female somatic maintenance or offspring. One is the nitrogenous reserves accumulated during their own larval feeding and stored in the fat body, which is usually depleted with egg development. Another is nectar feeding during the adult stage, although the nectar contains little nitrogen. The third resource is the contributions of the male ejaculate during copulation.

Nutrients Derived from Host Plants

For butterfly females, maximum egg production should correlate with the amount of stored nitrogenous nutrient available at eclosion, because reproductive products contain a high proportion of nitrogenous material in addition to carbohydrates and water (Karlsson 1996). Although the diet of most adults is nectar, consisting mainly of carbohydrates and water, Dunlap-Pianka et al. (1977) showed that adult intake of amino acids has affected the reproductive pattern in *Heliconius* spp. in which there is no evidence of ovarian senescence.

In most butterfly species, the difference in weight between a newborn caterpillar and the fully grown larva is more than 1000 fold. Most of that growth comes from what the caterpillar eats. The caterpillar needs to ensure that enough protein, fat, carbohydrates and vitamins are stored so that when it is ready to pupate it will be able to mature into a normal-sized adult, resulting in a fully functional adult eclosed. Accordingly, if the caterpillar does not eat quite enough, it will turn into a very small adult butterfly that is unlikely to have many offspring. If the caterpillar eats still less, then it may fail to complete its development. When food is unavailable in both quantity and quality, in addition a prolonged larval period for full growth must increase mortality risk by the predators because the larvae are generally immobile on the limited area of the host plants. Furthermore, because a long larval period results in late eclosion, particularly for male adults, they might not find virgin females late in the flying season. Therefore, feeding behaviour is a very serious business for the caterpillar, as stated by Vane-Wright (2003).

Egg production by adult females also requires protein-rich foods, the availability of which may critically constrain a female's lifetime reproductive success. Because the body size might be positively correlated with longevity, larval growth is primarily critical for females to increase offspring throughout the lifespan. Fat body tissue, representing reserves stored from the larval feeding and the site of yolk protein synthesis, fills the abdominal cavity of newly emerged females. Consequently, when there is no additional resource for the female to supplement larval-derived reserves, a relationship between reproductive reserves (body mass as an indicator) and reproductive output (the number of eggs actually laid) must appear. However, such relationship must become subtle in females of many polyandrous species. In the cabbage white, *Pieris rapae*, females that are reared on low-nitrogen diets reduced their allocation to the abdomen for increase of fecundity, whereas males maintain their allocation to reproductive structures, resulting in decline of spermatophore-producing glands (Tigreros et al. 2013).

Again, for the swallowtail butterfly, *Papilio xuthus*, in hilly regions without any agricultural citrus trees, leaves of short young trees of the host, *Zanthoxylum ailanthoides*, are preferred for oviposition by the females, and the larvae developed well on them, although the trees are never killed directly by the caterpillars (Watanabe 1983). The leaf biomass of the short trees is small and only sufficient for the full growth of a single larva. Consequently, all the leaves can be eaten by the larvae. However, such a short tree generally sprouts again about 2 weeks later, with seemingly little effect on continuous growth because it has the traits of a pioneer tree. Low population densities of the adult butterflies, and oviposition behaviour in the hilly region, suggest that only a single egg on a single short tree is common. In addition, relatively high mortalities in young larvae (Watanabe 1981) decrease the number of eggs or young larvae in the host trees. Therefore, even if more than two eggs are deposited on the same short tree, a single larva would develop well although the density does not increase.

Nutrients Derived from Cannibalism

In most papilionid and pierid butterflies, when it has escaped from its eggshell each caterpillar eats its own eggshell before eating the host plants (Fig. 4.9). The eggshell includes a carbohydrate-rich material as well as some key nutrients that females plaster on their eggs (Vane-Wright 2003). In some species, however, the earliest hatched caterpillar often devours unhatched eggs on the same leaf after eating its own eggshell (Rausher 1979). Although egg cannibalism is opportunistic, and regarded as abnormal, being prompted by such factors as extremely high egg densities and extreme shortage of food resources, a newly hatched caterpillar of the checkered white, *Pieris protodice*, frequently shows egg cannibalism for avoiding oviposition on the same leaf surface (Shapiro 1981). Watanabe and Yamaguchi (1993) also found that intra- and interspecific cannibalism among the

Fig. 4.9 A hatched caterpillar feeding on its own eggshell in the cabbage white, *Pieris rapae*

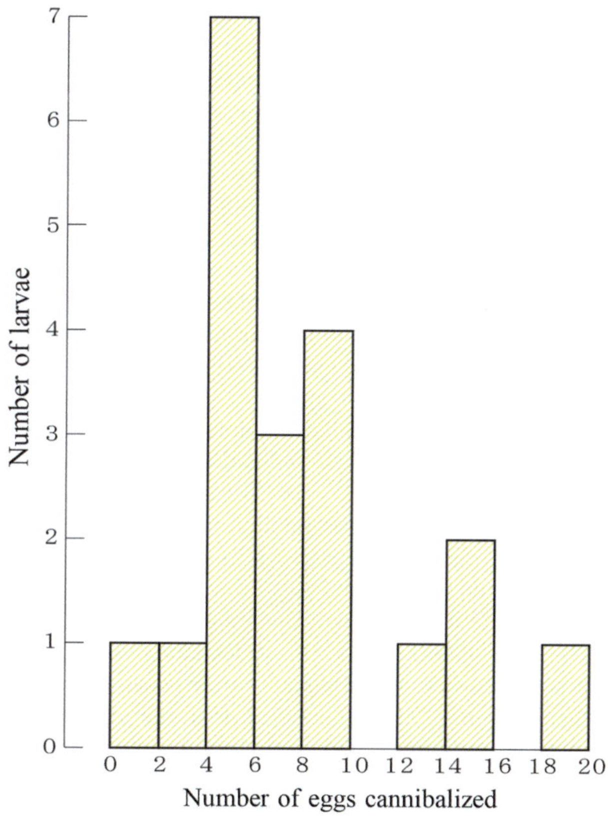

Fig. 4.10 Frequency distribution of the number of eggs cannibalized in the cabbage white, *Pieris rapae* (From Watanabe and Oh'ura 1997)

pierid butterflies, *Pieris rapae* and *Pieris melete*, involved eggs and newly hatched larvae on the same leaf of the host plant in the field.

Watanabe and Oh'ura (1997) experimentally clarified the egg cannibalism of the cabbage white, *Pieris rapae*, whose newly hatched caterpillars wander over the same leaf after eating their own eggshells as if seeming to search for other deposited eggs. Newly hatched caterpillars that are presented with conspecific eggs begin to eat the first egg 3 h after hatching, which is not significantly different from the starting time of feeding on leaves (for non-cannibals). The midgut remain yellowish brown for some time because the caterpillars have not yet fed on the green leaves. Thus, there is a delay before the start of feeding on the leaves. The midgut of cannibal larvae does not become greenish until 6.5 h after hatching. These cannibal larvae wander about, doing less feeding than non-cannibals.

The newly hatched caterpillar of the cabbage white, *Pieris rapae*, feeds on neighbouring eggs and then tries to feed on the next eggs (Fig. 4.10). Of 20 caterpillars, 1 feeds on 2 eggs during the first-instar stage and the other one feeds on 19 eggs, feeding on 8.3 eggs on average. Because the number of eggs offered experimentally as food is stable and they are constantly spaced, it can be seen from the data that cannibalism may not be a mortality factor in relationship to density of unhatched eggs.

Two kinds of egg cannibalism are seen in the cabbage white, *Pieris rapae*, as described by Watanabe and Oh'ura (1997): consumption of the entire egg, and nibbling of part of the eggshell. Both kinds of cannibalism are fatal to unhatched larvae. The former presumably provides the cannibals with nutrients. Eggs contain amino acids and organic compounds that are not directly derived from leaves. Such materials might facilitate larval development. In fact, the duration of the first-instar larvae of the cabbage white is significantly shorter for cannibals than for non-cannibals. However, there are no significant differences between cannibalising and non-cannibalising larvae over the whole larval period with regard to adult size and weight or female fecundity. On the other hand, the nibbling excludes unhatched conspecifics that would otherwise be competing for food. Therefore, possible advantages for the cannibal include nutrient gain as well as the elimination of possible competitors (Baur and Baur 1986).

Ohtani and Yamamoto (1985) found that female adults of the cabbage white, *Pieris rapae*, have no site fidelity, usually emigrating from their emergence site. Because the females lay eggs singly and seldom return to deposit on the same host plant, eggs on any one leaf may be deposited by two or more females. Porter (1992) stated that females laying a single egg distribute their eggs over a large number of host plants, and that this spread decreases the risks of predation and cannibalism among relatives and reduces competition with other larvae. Rothschild and Schoonhoven (1977) concluded that the cabbage white females discriminate between a cabbage leaf from which conspecific eggs had been removed and a clean control leaf. Few accounts of butterfly cannibalism have considered kin relatedness (Courtney 1986), although Jones (1982) reviewed cannibalism in relationship to kin selection.

Nutrients Derived from Males

Females have to increase the possibility of supplementing larval-derived resources with nutrients transferred during copulation, often referred to as 'nuptial gifts,' which function as paternal investment or as a mating effort or both. Ejaculate materials produced by accessory glands of the mated male are later absorbed within the female reproductive tract. After the termination of copulation in the swallowtail butterfly, *Papilio xuthus*, the weight of the spermatophore decreases slowly at first and then more rapidly (Fig. 4.11), and a considerable decline in weight is observed after 3 days (Watanabe et al. 2000). Then, 5 days after the copulation, the average spermatophore is one half of its initial weight, but the shape and colour are still oblate spherical and creamy, respectively. Thereafter, the spermatophore has been crushed, but the broken pieces of the capsule remained throughout her lifespan. Change in the decline of the spermatophore indicates that females do not use the ejaculates on the same day as the copulation. To improve body maintenance and

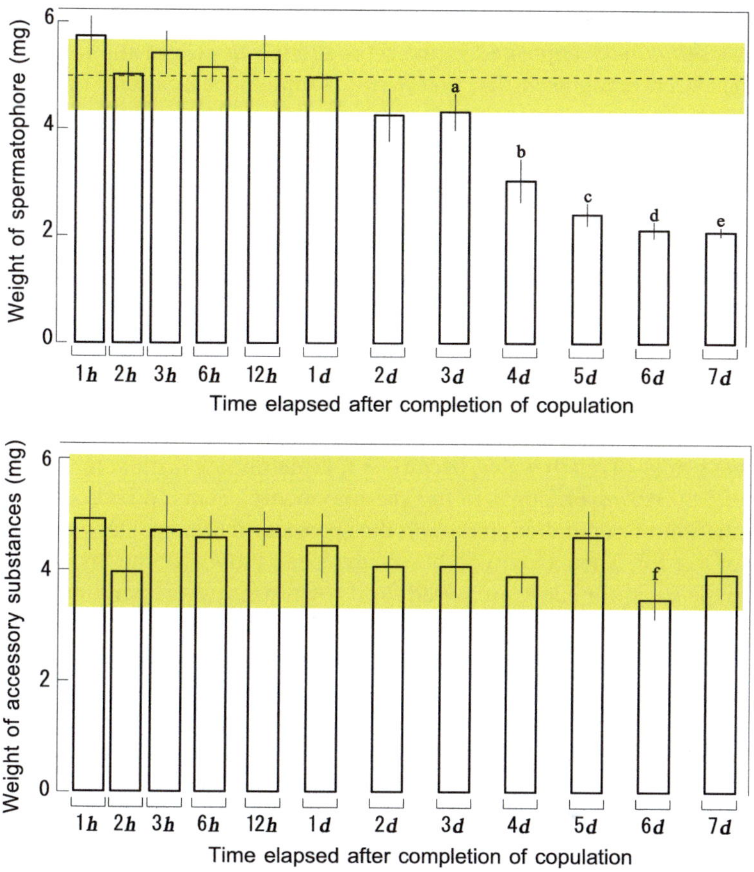

Fig. 4.11 Changes in the weight of ejaculates (accessory substances and spermatophore) after the completion of copulation in the swallowtail butterfly, *Papilio xuthus* (±SE). *Shaded areas* including a *dotted line* show SD and the average weight of ejaculates immediately after the completion of copulation. **a–f** show significant differences compared with the ejaculate immediately after completion of copulation (From Watanabe et al. 2000)

egg maturation, use of the male-derived nutrients must occur within 3 days after copulation. On the other hand, the weight of accessory substances declines more slowly than that of the spermatophore.

In addition to sperm transfer, male ejaculates contain many kinds of nutrients such as lipids (Marshall 1985), protein (Boggs and Gilbert 1979), sugar (Watanabe and Sato 1993), amino acids (Lai-Fook 1984), sterol (Marshall 1985), calcium phosphate (Oberhauser 1992), sodium (Pivnik and McNeil 1987) and zinc (Engebretson and Mason 1980). In the clouded sulphur, *Colias philodice*, ejaculates are a complex of proteins, hydrocarbons, triglycerides, diglycerides, sterols and phospholipids (Marshall 1982). Pivnik and McNeil (1987) demonstrated that males supplement the sodium requirements of the female via spermatophore transfer.

Lai-Fook (1991) showed that labelled phosphorus from the male canna skipper, *Calpodes ethlius*, was deposited in the reproductive tissues of the female. Therefore, it has been recognised that males may supplement a variety of the nutrient requirements of a female via spermatophore transfer. These nutrients are usually used for somatic maintenance of the female (Boggs and Watt 1981) and egg maturation in the ovaries (Watanabe 1988).

Watanabe and Sato (1993) examined the spermatophore in the field-captured monogamous females of the cabbage white, *Pieris rapae*, of age FF, and showed that the size was not significantly different from that of the females immediately after copulation in the laboratory, whereas the spermatophores in the female of age F and B are smaller than those either in laboratory females or field-captured females of age FF, indicating that females gradually absorbed the spermatophore (Fig. 4.12). In addition, significant quantities of sugar were contained in the spermatophore immediately after copulation in the laboratory experiment. Such sugar must be dissolved in water, because a spermatophore is more than 80 % water (Boggs 1981). However, sugar in the spermatophore seems to be consumed upon the termination of copulation, although the spermatophore size does not change in females of age FF. Then, two hypotheses have been proposed: the first is that sperm could use the sugar for energy required for its own survival or sperm migration, and the second is that females could be supplied with the energy for activity in the bursa copulatrix, although the sugar might not contribute nutritionally to female somatic maintenance or egg production.

Radiolabelled amino acids from the spermatophore have been demonstrated to be incorporated into female somatic tissue and eggs in the monarch butterfly, *Danaus plexippus*, and the longwings, *Heliconius* spp. (Boggs and Gilbert 1979),

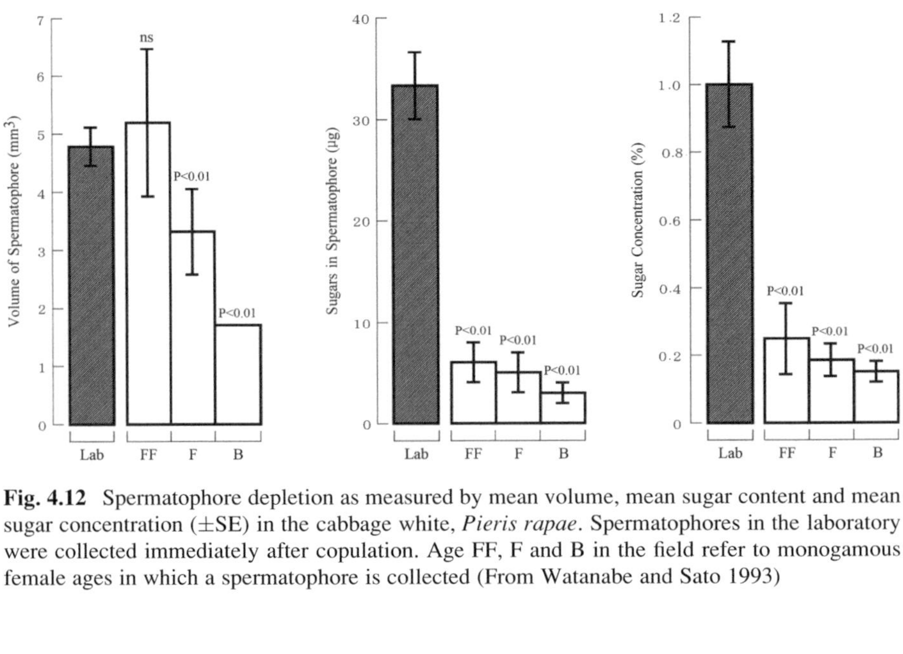

Fig. 4.12 Spermatophore depletion as measured by mean volume, mean sugar content and mean sugar concentration (±SE) in the cabbage white, *Pieris rapae*. Spermatophores in the laboratory were collected immediately after copulation. Age FF, F and B in the field refer to monogamous female ages in which a spermatophore is collected (From Watanabe and Sato 1993)

the orange sulphur, *Colias eurytheme* (Boggs and Watt 1981), the green-veined white, *Pieris napi* (Wiklund et al. 1993) and others. In the orange sulphur, incorporation of male-derived nutrients into eggs is rapid, the day after mating. For the green-veined white, the rate of incorporation of the nutrients into eggs peaked at 3 to 4 days following mating. On the other hand, in longwings the peak of incorporation of male-derived amino acids into eggs varies from 3 to as many as 14 days after mating. Therefore, the eggs laid by the female contained the radioactive label in the many species with different mating and different adult feeding habits, that is, the polyandrous females of the species which feed on nectar and show high multiple matings, the polyandrous females of the species that feed on pollen as well as nectar, and even the monandrous females of the species. However, Svärd and Wiklund (1991) noticed a considerable time lag of eventual incorporation of male-derived nutrients into eggs in the monandrous species, where females eclose with abundant mature eggs ready to be laid soon after mating.

The female may consume the nuptial gift primarily to gain nutrients, and there can be seen a passive consequence of the fact that larger ejaculates take longer to consume. Because of gradually absorbing the spermatophore after copulation, the stage of spermatophore collapse in the bursa copulatrix can be used to determine the relative length of the time since the spermatophore was transferred from the male (Konagaya and Watanabe 2015). The spermatophore found in the bursa copulatrix of wild females could be classified into one of four stages on the basis of its shape to assess when the mating occurred. As shown in Fig. 4.13, for the common grass yellow, *Eurema mandarina*, a spermatophore at stage I is defined as an intact one that may have been recently transferred by a male and not yet depleted by the female. A spermatophore at stage II is spherical without the neck. Because the spermatophore body was intact, it might also have been derived from a male with which the female had recently mated. A spermatophore at stage III has a concave shape, like a bowl, which resulted from female absorption. A spermatophore at the stage IV is largely destroyed and too small to weigh, as the weight was less than 10 μg in the case of the common grass yellow, suggesting that a considerably long time had passed since the mating.

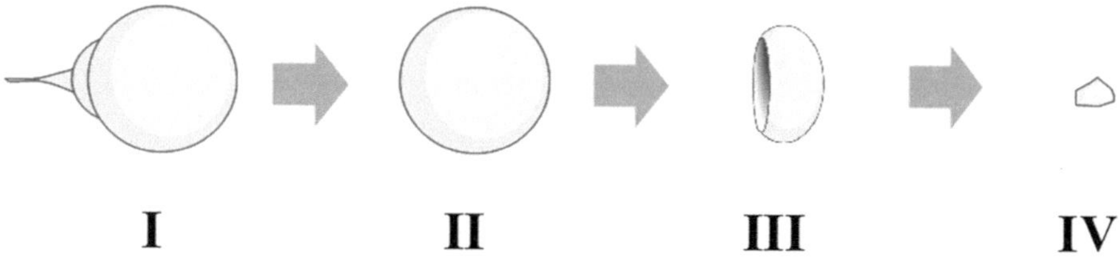

Fig. 4.13 Schematic representation of the process of spermatophore collapse in the bursa copulatrix of *Eurema mandarina* females. *Stage I* shows the intact spermatophore immediately after copulation. The spermatophore losing its neck is in *stage II*. Probably by the activity of the bursa copulatrix, the spermatophore is squashed and the nutrients it contains are absorbed by the female (*stage III*). Consequently, only a tiny piece of the spermatophore capsule remains in the bursa copulatrix (*stage IV*) (From Konagaya and Watanabe 2015)

The potential importance of male-derived nutrients might correlate positively with the size of the male ejaculate (Svärd and Wiklund 1989). A demonstration that the nuptial gifts of males represent paternal investment requires that the gift results in an increase in offspring fitness and/or the number, inseminated by the male (Vahed 1998). Thus, the nuptial gifts have to be attractive to females, with some nutritional value. Nutrients provided by the spermatophore transferred from the male have been incorporated into eggs in most butterfly species. A single spermatophore produced by males of the green-veined white, *Pieris napi*, contains the nitrogen equivalent of a maximum of 70 eggs (Karlsson 1998).

Large spermatophores appear to be maintained by selection of the male to maximise ejaculate transfer and thereby counter the effects of sperm competition. Svärd and Wiklund (1986) showed that ejaculate production capacity of males increase with degree of polyandry across the species. Males of the polyandrous swallowtail butterfly, *Papilio xuthus*, can replenish ejaculate materials, using not only the fat body derived from larval feeding but also sugars derived from adult feeding (Watanabe and Hirota 1999). Svärd and Wiklund (1989) stated that males belonging to polyandrous species were able to transfer a second ejaculate as large as the first ejaculate within a couple of days. Recovery time for males in monandrous species to produce a second ejaculate as large as the first one might be longer than that for males in polyandrous species.

In addition to the increase in ejaculate size with degree of polyandry across the species, the content in the ejaculates is more protein rich when the degree of polyandry increases (Bissoondath and Wiklund 1995) and seems to be associated with the nitrogen content of male abdomens. In fact, among males, the amount of abdomen nitrogen increases with the degree of polyandry (Fig. 4.14), whereas thorax nitrogen, abdomen carbon and thorax carbon are not influenced by the

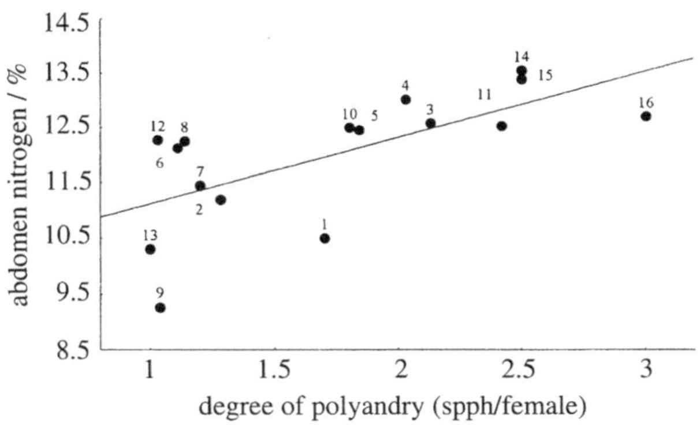

Fig. 4.14 Relationship between degree of polyandry (mean number of spermatophores per female) and abdomen nitrogen. $r = 0.662$, $p = 0.005$. *1 Aporia crataegi, 2 Pieris brassicae, 3 Pieris rapae, 4 Pieris napi, 5 Pontia daplidice, 6 Anthochalis cardamines, 7 Gonepteryx rhamni, 8 Leptidae sinapis, 9 Pararge aegeria, 10 Aglais urticae, 11 Polygonia c-album, 12 Speyeria mormonia, 13 Heliconius charitonius, 14 Heliconius melpomene, 15 Heliconius pachinus, 16 Heliconius hecale* (From Karlsson 1996)

degree of polyandry. Therefore, Karlsson (1996) concluded that males of polyandrous species produce heavier ejaculates with higher nutritional quality than do males of monandrous species.

Females that received heavier spermatophores or a larger number of them showed greater longevity as well as higher reproductive output in the green-veined white, *Pieris napi* (Wiklund et al. 1993) and the alfalfa butterfly, *Colias eurytheme* (Rutowski et al. 1987). Prudic et al. (2011) suggested that females of the squinting bush brown, *Bicyclus anynana*, in their dry seasonal form prolonged their longevity using nuptial gifts from males. Females of the monarch butterfly, *Danaus plexippus*, might use nutrients in the spermatophore during the overwintering period (Leong et al. 2012). Nuptial gifts to overwintering females are believed to be beneficial for males with which they had mated before overwintering, because the "overwintering success of sperm" completely depends on the mortality of females (see Chap. 7, this volume).

For the females, the reproductive value of multiple matings has been discussed as a remedy for an insufficient supply of viable sperm resulting from an initial mating: for the enhancement of genetic diversity among offspring, and for the facilitation of a paternal nutritional investment. Because females deposit eggs during the time between each mating, genetic diversity among her offspring may be increased accordingly. Suzuki (1979) pointed out, however, that there are some difficulties in measuring the selective advantage of such genetic diversity under natural conditions.

Multiple Mating by Females

Proposed evolutionary explanations for the observed mating numbers in polyandrous females might include increasing the genetic diversity of mates, achieving an adequate sperm supply, enhancing female survival and reproductive output and avoiding the time costs associated with the mating of insistent males. For the number of matings to be a character under selection, it must have primarily a genetic basis. However, positive correlations between mating frequency and population density are reported in a few butterfly species. Odendaal et al. (1985) showed that females in those species exhibit little or no choice of mate, suggesting that at high density the competition among males for females becomes more intense and the number of matings in females increases. Because of a male-killing *Wolbachia* infection, local populations of the great eggfly butterfly, *Hypolimnas bolina*, in South Pacific Islands are female biased (Charlat et al. 2007). Because of the low densities of surviving males, those males have to increase the number of matings, which results in decreased spermatophore mass because male nutrients for copulations are limited. Then, females are apt to re-mate because they received such a small spermatophore in each copulation. Nonetheless, in general, the rate of increase in the number of matings with age was not different from the fluctuation of

Fig. 4.15 Male of the Asian green-veined white, *Pieris melete*, feeding on red clover

Table 4.1 Frequency distribution and the mean number of spermatophores in wild-caught females of the Asian green-veined white, *Pieris melete*, for five age classes (FF, F, B, BB and BBB)

	Age class				
Number of spermatophores contained	FF	F	B	BB	BBB
0	4	0	0	0	0
1	8	6	4	3	0
2	0	9	11	11	5
3	0	1	15	16	4
4	0	0	5	1	14
5	0	0	0	1	1
6	0	0	1	0	1
Mean number of spermatophores	0.7	1.7	2.7	2.6	3.6

Source: Ando and Watanabe (1993)

male density and the number of active males searching for mates (Watanabe and Ando 1993).

In the Asian green-veined white, *Pieris melete* (Fig. 4.15), Ando and Watanabe (1993) examined the number of spermatophores in the bursa copulatrix of wild-caught females in relationship to age and estimated that about four matings occurred in their lifetime (Table 4.1). In swallowtail butterflies (Fig. 4.16), the older females of *Papilio xuthus* in the wild mated a maximum of five times, whereas *Papilio machaon* females seem to be monandrous (Watanabe and Nozato 1986). It was reported that pushing previously stored spermatophores deep in the bursa copulatrix is advantageous for the last male to inseminate his sperm (Stern and Smith 1960). The order of deposition of spermatophores could be determined by the relative position of the spermatophores and probably by the order of depletion in many butterfly species (Rutowski et al. 1981).

Female butterflies might be able to make a more accurate assessment of male quality after copulation (Wedell and Cook 1998). In the green-veined white, *Pieris*

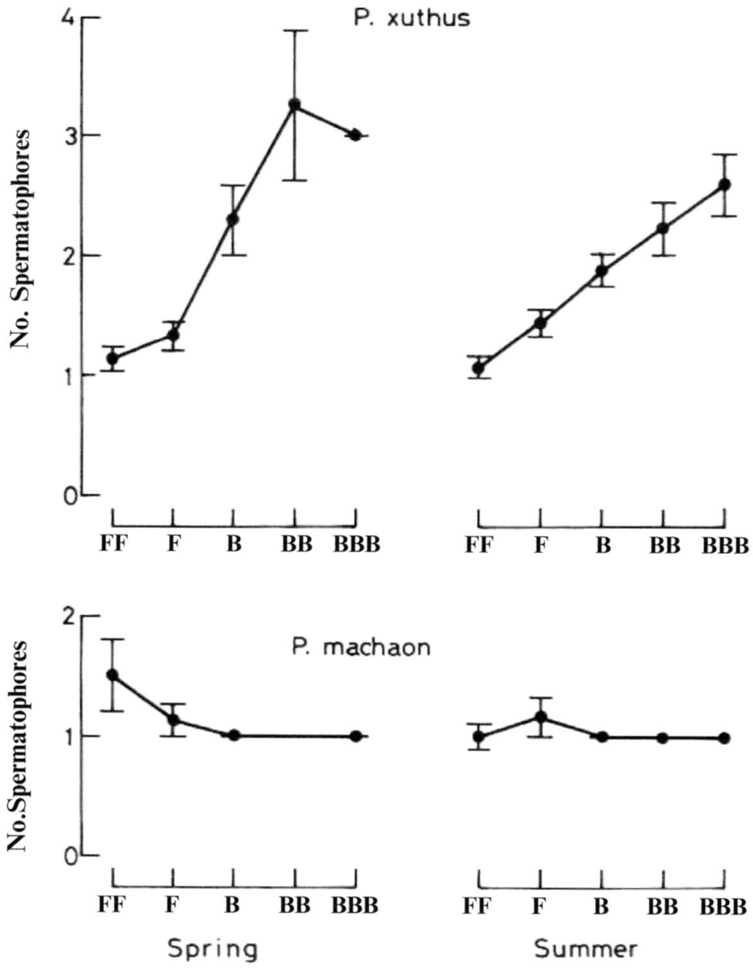

Fig. 4.16 Change in number of spermatophores found in the bursa copulatrix of the field-captured swallowtail butterflies, *Papilio xuthus* and *Papilio machaon*, of respective ages in both spring and summer generations (From Watanabe and Nozato 1986)

napi, females appear unable to detect male mating status and are just as likely to mate with already mated males, although they will receive smaller nutrient donations, resulting in lower fecundity (Wiklund et al. 1998). Välimäki et al. (2011) also showed that females of the green-veined white do not discriminate against close relatives regardless of the initial possibility to evaluate and choose their mates and so avoid the costs of inbreeding. In Finland, the potential mates of the green-veined white are encountered sequentially, and thus even the probability of encounters between the potential mates would be relatively high. However, the precopulatory kin-recognition mechanism may not be strongly selected. Instead, the female re-mating interval is sensitive to mating type, so that mating with a kin resulted

Table 4.2 Frequency distributions of the number of spermatophores in the bursa copulatrix of females in two morphs of the pale clouded yellow, *Colias erate*, captured in the field. Number of matings increased with age for either morph (Kendall test at $p < 0.05$)

		Age class				
Morph	Number of spermatophores	FF	F	B	BB	BBB
Alba	0	5	0	0	0	0
	1	143	79	48	10	1
	2	22	28	51	34	13
	3	0	3	6	15	2
	4	0	0	0	3	2
	Mean	1.13	1.30	1.60	2.05[a]	2.27[b]
Andromorph	0	0	0	0	0	0
	1	43	28	21	5	3
	2	5	12	29	13	7
	3	0	0	1	2	0
	Mean	1.10	1.32	1.63	1.80[a]	1.70[b]

[a, b]Mann–Whitney *U* test at $p < 0.05$
Source: Nakanishi et al. (1996)

in relatively short re-mating intervals with a small number of eggs fertilised by the relatives under the last male precedence.

In the case of the pale clouded yellow, *Colias erate*, in which females show wing dimorphism, Nakanishi et al. (1996) observed that most young albas (FF) had a single spermatophore in the bursa copulatrix, resulting in an average mating number of 1.1 (Table 4.2). The number of matings for alba increased with age, and the lifetime mating number exceeded 2. However, the mating frequency of the andromorph is lower than that of the alba by age BB and BBB. In addition, every female carried a decreased load of immature eggs within her lifetime. For monogamous females, 550 and 400 eggs are laid by andromorphs and albas, respectively, although virgin females of either morph contain about 750 immature eggs. In contrast, among polygamous females, the lifetime number of eggs in andromorph and alba is 550 and 650, respectively, suggesting that old albas are apt to accept further mating more easily than the old andromorphs, from the aspect of increase in the number of eggs laid throughout the lifespan. Indeed, Watanabe et al. (1997) showed that males persistently courted albas and sometimes harassed copulating pairs involving alba rather than andromorph in the field.

An increase in fecundity resulting from the receipt of a larger spermatophore or a larger number of spermatophores through multiple mating might be caused by the effects of extra sperm or oviposition stimulants in the ejaculates: the former is to achieve an adequate sperm supply. Labine (1966) pointed out that sperm may deteriorate with age after insemination and must be replaced with a fresh supply, because there may be too few spermatozoa in one insemination to fertilise all the eggs that a female will lay. Lederhouse (1981) pointed out that the benefits of multiple matings can be a substantial effect on increasing good sperm supply.

However, as a rule one mating provides a female in many butterfly species with enough spermatozoa to fertilise all her eggs (Rutowski 1984). Thus, polyandry might not be ascribed simply to sperm replenishment.

The effects of multiple matings are obviously of great importance for understanding the selective advantage of behaviour for participating males. In the swallowtail butterfly, *Papilio xuthus*, a mating 1 day after emergence led to an increase in the numbers of mature eggs, whereas that a few days after emergence does not (Watanabe 1988). No females which were not allowed to oviposit also had increased numbers of mature eggs after the second or third mating. This observation suggests that certain afferent impulses enhance the activity of egg production after copulation, but such impulses do not occur later than 1 day from emergence. Hormonal stimulation of egg production might be elicited by the male.

Sugawara (1979) suggested that the impulses are induced by the presence of a spermatophore, triggering the mate refusal posture in the female cabbage white, *Pieris rapae*. Therefore, if a female does not mate and feeds exclusively on nectar, she might lose her ability to produce extra eggs in response to mating as time goes on. Thus, females are time limited because the timing of copulation may affect reproductive success. Watanabe and Nozato (1986) pointed out that the swallowtail butterfly, *Papilio xuthus*, mated upon emergence. It is generally accepted that most female butterflies have mated on emergence or within a day afterwards (Labine 1968). Females unable to mate at emergence because of bad weather might actively seek males for mating. Indeed, Rutowski (1980) reported that the female of the sulphur butterflies, *Colias* spp., chased males after she had exploited the inseminated sperm.

Because spermatophores contain nutrition such as proteins and are used by females for egg maturation and somatic maintenance, the number of spermatophores must affect the reproductive output of females in polyandrous species. South and Lewis (2011) reviewed the influence of male ejaculate quantity that contains numerous substances produced by male accessory glands as well as sperm on female fitness during copulation. A possible relationship between mating frequency and lifetime fecundity may exist in many butterfly species (Boggs and Watt 1981). The positive effect of multiple mating on female fecundity has been found in the swallowtail butterfly, *Papilio xuthus* (Watanabe 1988), the green veined white, *Pieris napi* (Wiklund et al. 1993), the cabbage white, *Pieris rapae* (Watanabe and Ando 1993) and so on, although Svärd and Wiklund (1988) found no effect of the number of matings on female fecundity. On the other hand, the females of the monandrous pipevine swallowtail, *Battus philenor*, show little effect of spermatophore nutrition on reproductive output (Konagaya et al. 2015). Alternatively, larger females show higher potential fecundity, indicating that the reproductive output depended on female size rather than the number of matings in monandrous species.

Extensive studies have been made of the reproductive output of the pierid butterflies (Stern and Smith 1960). Data on lifetime reproductive schedule in females have been proposed for the cabbage white, *Pieris rapae* (Suzuki 1978), the Asian green veined white, *Pieris melete* (Kimura and Tsubaki 1986) and the green veined white, *Pieris napi* (Yamamoto 1981). The effect of mating frequency on oviposition, however, has not been involved in such studies, because all the females used in the laboratory were artificially monogamous.

The weight of an egg laid by females mated only once decreased with female age in the cabbage white, *Pieris rapae* (Kimura and Tsubaki 1985). If every female has a similar quantity of nutrients derived from larval food plants, egg size in old females must tend to decrease. In fact, mature eggs in the ovaries of young monogamous females were larger than those of older monogamous females (Watanabe and Ando 1994), although there was no significant difference in that of polygamous females, indicating that the mature eggs of old polygamous females seemed to be larger than that of monogamous females. Therefore, the variation in egg size laid by females within each butterfly species might be associated with the number of spermatophores.

Figure 4.17 shows the lifetime changes in the number of immature eggs in relationship to mating frequency for the cabbage white, *Pieris rapae*. Virgin females load about 450 immature eggs. When a female mated once, the number of immature eggs decreased to 400. The remainder (50 immature eggs) were likely to develop

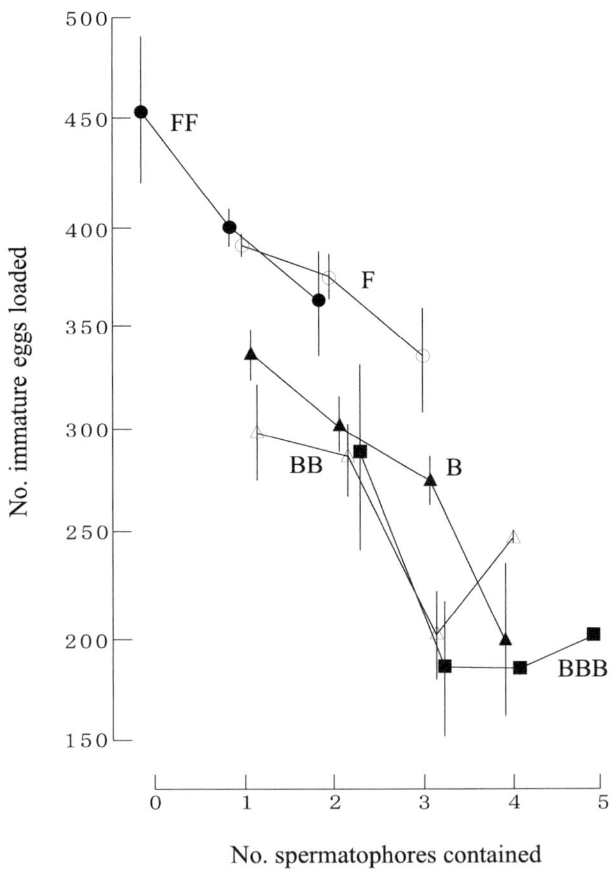

Fig. 4.17 Change in number of immature eggs loaded for respective age class in relationship to mating frequency indicated by number of spermatophores contained in the bursa copulatrix of wild-caught females of the cabbage white, *Pieris rapae*. Each *bar* represents SE (From Watanabe and Ando 1993)

into mature eggs to be laid. The number of immature eggs decreased with the increase in mating frequency in each age class. Because fewer than 100 mature and submature eggs are loaded in most females, decrease in immature eggs resulted from oviposition. Because a virgin female had 450 immature eggs, a monogamous female of age BB may have laid 150 eggs. On the other hand, the oldest female (age BBB) that mated more than three times loaded 200 immature eggs, suggesting that she had laid 250 eggs.

Wiklund et al. (1993) also showed that polygamous females of the green-veined white, *Pieris napi*, had greater lifetime fecundity than monogamous females, laying on average 1.61 times as many eggs. This finding suggests that one of the benefits from polyandry may be to enhance female reproductive output through male nutrient delivery. However, Välimäki et al. (2006) found that in the green-veined white inhabiting Sweden, females with a low mating rate produced more eggs during the early days of reproduction than comparable females with a high mating rate; this might indicate that foregoing the chance of obtaining nutrients through multiple matings is beneficial for a female in terms of egg quantity, if the time for reproduction is limited. Then, the benefits of multiple mating arose about 1 week after the first mating, by which time the reproductive rate of females with a high mating rate had exceeded that of females with a low mating rate.

Watanabe (1988) showed, in the swallowtail butterfly, *Papilio xuthus*, that the number of eggs deposited by the female mated once was high on the day after mating and then decreased (Fig. 4.18). The egg diameter was constant throughout the oviposition period (about 1.24 mm). In this experiment, all the females were kept from oviposition on the 6th day after emergence. Twenty-two of the 37 females recommenced oviposition after the 7th day. However, the number of eggs deposited did not regain the earlier levels. Most of them continued ovipositing until the 11th

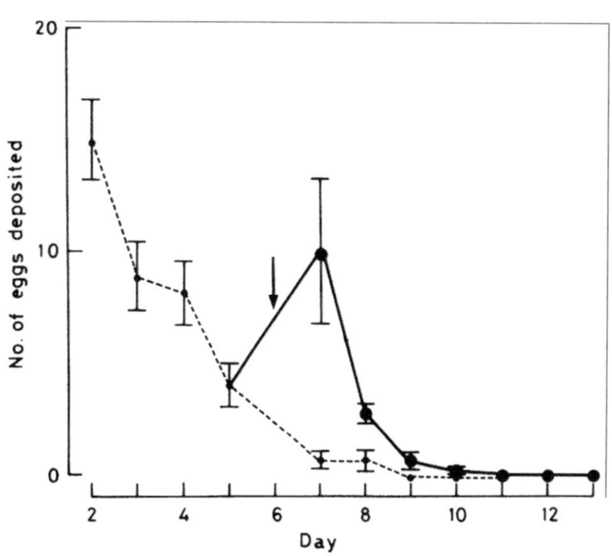

Fig. 4.18 Change in number of eggs deposited by a female with a single mating (*broken line*) and with a double mating (*solid line*) in the swallowtail butterfly, *Papilio xuthus*. All females were mated at 1 day old, the day after emergence. *Arrow* indicates day of re-mating for 15 of 37 females mated. The other 22 females were kept from oviposition on that daEy (From Watanabe 1988)

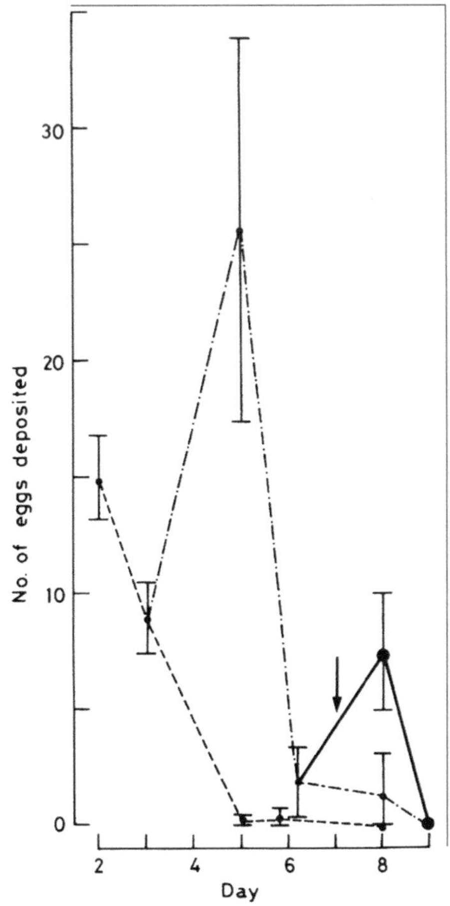

Fig. 4.19 Change in number of eggs deposited by a female with a single mating (*broken line*), with a double mating (*dashed* and *dotted line*) and with a triple mating (*solid line*) in the swallowtail butterfly, *Papilio xuthus*. All females were mated at 1 day old, the day after emergence. Nine of 13 females mated were re-mated on the 4th day. *Arrow* indicates day of third mating for 6 of 9 double-mated females (From Watanabe 1988)

day. In contrast, 15 of 37 females re-mated on the 6th day after emergence. These double-mated females deposited significantly more eggs on the 7th and the 8th day than did single-mated females. However, the number of eggs deposited on the 9th day was not significantly different between single- and double-mating females.

Figure 4.19 shows the change in the number of eggs deposited by a single-, a double-, and a triple-mated female of the swallowtail butterfly, *Papilio xuthus*. Thirteen single-mated females paused in their oviposition on the 4th day after emergence. Then, 4 of 13 females oviposited further after the 5th day. The other 9 females re-mated on the 4th day, and then oviposited after the 5th day. Again, oviposition paused on the 7th day, when 6 of 9 females mated yet again. All oviposited after the 8th day. The number of eggs deposited on the day after the second mating showed an increase over that of single-mated females on the same day. Furthermore, when such females had been re-mated again on the 7th day, the

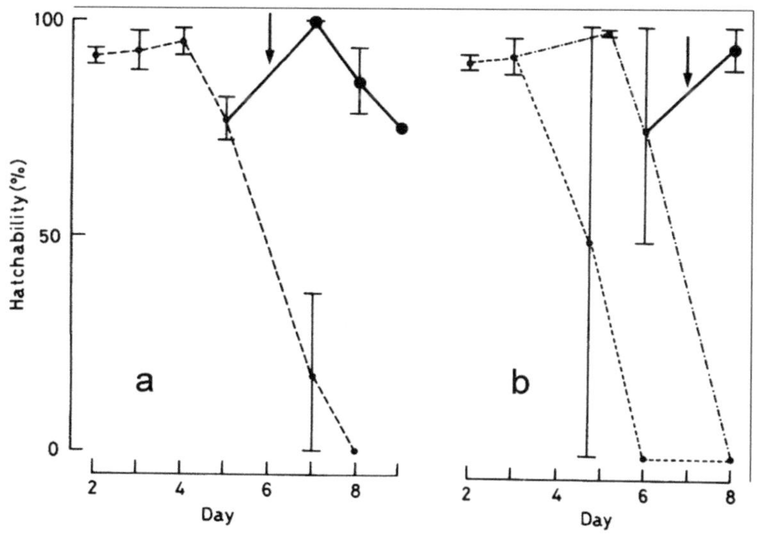

Fig. 4.20 Change in hatchability of eggs deposited by females of the swallowtail butterfly, *Papilio xuthus*: a double mating (*solid line*) compared with a single mating (*broken line*) (**a**); a triple mating (*solid line*) compared with a double mating (*dashed* and *dotted line*) and a single mating (*broken line*) (**b**). *Arrows* indicate the days of the second (**a**) and the third (**b**) mating. Numbers of eggs examined were derived from those deposited by females in Figs. 4.17 and 4.18 (From Watanabe 1988)

number of eggs deposited on the 8th day showed an increase compared with other females.

The mean incubation period of eggs of the swallowtail butterfly, *Papilio xuthus*, is about 3 days. A high hatchability was observed in eggs from 3-day-old single-mated females but it decreased rapidly in eggs from older, single-mated females (Fig. 4.20). Although females were kept from oviposition on the 4th and the 6th day, hatchability did not recover in eggs deposited. When the females re-mated at the 6th day, the hatchability of eggs produced on the 7th and 8th days was higher than those of single-mated females. This recovery in hatchability was also found in the 5- and 6-day-old females which paused from oviposition on the 4th day. However, in this case hatchability gradually declined. The hatchability in eggs deposited by triple-mated females recovered following the third mating. Therefore, multiple matings increase the reproductive output of the females.

References

Ando S, Watanabe M (1993) Mating frequency and egg load in the white butterfly, *Pieris melete* Menetries, in a wild environment. Jpn J Ecol 43:111–114. (in Japanese)

Baur B, Baur A (1986) Proximate factors influencing egg cannibalism in the land snail *Arianta arbustorium* (Pulmonata, Helicidae). Oecologia (Berl) 70:283–287

Bissoondath CJ, Wiklund C (1995) Protein content of spermatophores in relation to monandry/polyandry in butterflies. Behav Ecol Sociobiol 37:365–371
Boggs CL (1981) Selection pressures affecting male nutrient investment at mating in heliconiine butterflies. Evolution 35:931–940
Boggs CL, Gilbert LE (1979) Male contribution to egg production in butterflies: evidence for transfer of nutrients at mating. Science 206:83–84
Boggs CL, Watt WB (1981) Population structure of pierid butterflies. IV. Genetic and physiological investment in offspring by male *Colias*. Oecologia (Berl) 50:320–324
Charlat S, Reuter M, Dyson EA, Hoenett EA, Duplouy A, Davies N, Roderick GK, Wedell N, Hurst GDD (2007) Male-killing bacteria trigger a cycle of increasing male fatigue and female promiscuity. Curr Biol 17:273–277
Courtney SP (1986) The ecology of pierid butterflies: dynamics and interactions. Adv Ecol Res 15:51–131
Dunlap-Pianka H, Boggs CL, Gilbert LE (1977) Ovarian dynamics in heliconiine butterflies: programmed senescence versus eternal youth. Science 197:487–490
Engebretson JA, Mason WH (1980) Transfer of ^{65}Zn at mating in *Heliothis virescens*. Environ Entomol 9:119–121
Feeny P (1970) Seasonal changes in oak leaf tannins and nutrients as a cause of spring feeding by winter moth caterpillars. Ecology 51:565–581
Feltwell J (1986) The natural history of butterflies. Croom Helm, London
Honda K, Hayashi N (1995) Chemical factors in rutaceous plants regulating host selection by two swallowtail butterflies, *Papilio protenor* and *P. xuthus* (Lepidoptera: Papilionidae). Appl Entomol Zool 30:327–334
Jones JS (1982) Of cannibals and kin. Nature (Lond) 299:202–203
Karlsson B. (1996) Male reproductive reserves in relation to mating system in butterflies: a comparative study. Proc R Soc Lond B 263:187–192
Karlsson B (1998) Nuptial gifts, resource budgets, and reproductive output in a polyandrous butterfly. Ecology 79:2931–2940
Kimura K, Tsubaki Y (1985) Egg weight variation associated with female age in *Pieris rapae crucivora* Boisduval (Lepidoptera: Pieridae). Appl Entomol Zool 20:500–501
Kimura K, Tsubaki Y (1986) Oviposition schedule of the small white butterfly, *Pieris melete* Menetries (Lepidoptera: Pieridae). Appl Entomol Zool 21:28–32
Konagaya T, Watanabe M (2015) Adaptive significance of the mating of autumn-morph females with non-overwintering summer-morph males in the Japanese common grass yellow, *Eurema mandarina* (Lepidoptera: Pieridae). Appl Entomol Zool 50:41–47
Konagaya T, Mutoh N, Suzuki M, Rutowski RL, Watanabe M (2015) Estimates of female lifetime fecundity and changes in the number and types of sperm stored with age and time since mating in the monandrous swallowtail butterfly, *Battus philenor* (Lepidoptera: Papilionidae) in the Arizona desert. Appl Entomol Zool 50:311–316
Labine PA (1966) The population biology of the butterfly, *Euphydryas editha*. IV. Sperm precedence: a preliminary report. Evolution 20:580–586
Labine PA (1968) The population biology of the butterfly, *Euphydryas editha*. VIII. Oviposition and its relation to patterns of oviposition in other butterflies. Evolution 22:799–805
Lai-Fook J (1984) The spermatophore of the skipper, *Calpodes ethlius* (Hesperiidae: Lepidoptera): the sperm sac. Can J Zool 62:1135–1143
Lai-Fook J (1991) Absorption of phosphorus from the spermatophore through the cuticle of the bursa copulatrix of the butterfly, *Calpodes ethilus*. Tissue Cell 23:247–259
Lederhouse RC (1981) The effect of female mating frequency on egg fertility in the black swallowtail, *Papilio polyxenes asterius* (Papilionidae). J Lepid Soc 35:266–277
Leong KHL, Yoshimura MA, Williams C (2012) Adaptive significance of previously mated monarch butterfly females (*Danaus plexippus* (Linnaeus)) overwintering at a California winter site. J Lepid Soc 66:205–210

Marshall LD (1982) Male nutrient investment in the Lepidoptera: what nutrients should males invest? Am Nat 120:273–279

Marshall LD (1985) Protein and lipid composition of *Colias philodice* and *C. eurytheme* spermatophores and their changes over time. J Res Lepid 24:21–30

Nakanishi Y, Watanabe M, Ito T (1996) Differences in lifetime reproductive output and mating frequency of two female morphs of the sulfur butterfly, *Colias erate* (Lepidoptera: Pieridae). J Res Lepid 35:1–8

Oberhauser KS (1992) Rate of ejaculate breakdown and intermating intervals in monarch butterflies. Behav Ecol Sociobiol 31:367–373

Odendaal FJ, Iwasa Y, Ehrlich PR (1985) Duration of female availability and its effect on butterfly mating systems. Am Nat 125:673–678

Ohsaki N (1980) Comparative population studies of three *Pieris* butterflies, *P. rapae*, *P. melete* and *P. napi*, living in the same area. II. Utilization of patchy habitats by adults through migratory and non-migratory movements. Res Popul Ecol 22:163–183

Ohtani T, Yamamoto M (1985) The adult behavior of the Japanese cabbage white (Lepidoptera, Pieridae) in the field. II. Ecological aspects of major behavior patterns. Tyo to Ga 36:43–76

Porter K (1992) Eggs and egg-laying. In: Dennis RHD (ed) The ecology of butterflies in Britain. Oxford University Press, Oxford, pp 46–72

Prudic KL, Jeon C, Cao H, Monteiro A (2011) Developmental plasticity in sexual roles of butterfly species drives mutual sexual ornamentation. Science 331:73–75

Rausher MD (1978) Search image for leaf shape in a butterfly. Science 200:1071–1073

Rausher MD (1979) Egg recognition: its advantage to a butterfly. Anim Behav 27:1034–1040

Rothschild M, Schoonhoven LM (1977) Assessment of egg load by *Pieris brassicae* (Lep., Pieridae). Nature (Lond) 266:352–355

Rutowski RL (1980) Male scent-producing structures in *Colias* butterflies—function, localization, and features. J Chem Ecol 6:13–26

Rutowski RL (1984) Sexual selection and the evolution of butterfly mating behavior. J Res Lepid 23:125–142

Rutowski RL, Long CE, Marshall LD, Vetter RS (1981) Courtship solicitation by *Colias* females (Lepidoptera: Pieridae). Am Midl Nat 105:334–340

Rutowski RL, Gilchrist GW, Terkanian B (1987) Female butterflies mated with recently mated males show reduced reproductive output. Behav Ecol Sociobiol 20:319–322

Shapiro AM (1970) The role of sexual behavior in density-related dispersal of pierid butterflies. Am Nat 104:367–372

Shapiro AM (1981) The pierid red-egg syndrome. Am Nat 117:276–294

Sharifi S, Zarea N (1970) Biology of the citrus butterfly *Papilio demoleus demoleus* (Lepidoptera: Papilionidae). Ann Entomol Soc Am 63:1211–1213

South A, Lewis SM (2011) The influence of male ejaculate quantity on female fitness: a meta-analysis. Biol Rev 86:299–309

Stern VM, Smith RF (1960) Factors affecting egg production and oviposition in populations of *Colias philodice eurytheme* Boisduval (Lepidoptera: Pieridae). Hilgardia 29:411–454

Sugawara T (1979) Stretch reception in the bursa copulatrix of the butterfly, *Pieris rapae crucivora*, and its role in behaviour. J Comp Physiol 130:191–199

Suzuki Y (1978) Adult longevity and reproductive potential of the small cabbage white, *Pieris rapae crucivora* Boisudual (Lepidoptera: Pieridae). Appl Entomol Zool 13:312–313

Suzuki Y (1979) Mating frequency in females of the small cabbage white, *Pieris rapae crucivora* Boisduval (Lepidoptera: Pieridae). Kontyu 47:335–339

Svärd L, Wiklund C (1986) Different ejaculate strategies in first versus subsequent matings in the monarch butterfly *Papilio machaon* L. Behav Ecol Sociobiol 18:325–330

Svärd L, Wiklund C (1988) Prolonged mating in the monarch butterfly *Danaus plexippus* and nightfall as a cue for sperm transfer. Oikos 52:351–354

Svärd L, Wiklund C (1989) Mass and production rate of ejaculates in relation to monandry/polyandry in butterflies. Behav Ecol Sociobiol 24:395–402

Svärd L, Wiklund C (1991) The effect of ejaculate mass on female reproductive output in the European swallowtail butterfly, *Papilio machaon* (L.) (Lepidoptera: Papilionidae). J Insect Behav 4:33–41
Tigreros N, Sass EM, Lewis SM (2013) Sex-specific response to nutrient limitation and its effects on female mating success in a gift-giving butterfly. Evol Ecol 27:1145–1158
Vahed K (1998) The function of nuptial feeding in insects: a review of empirical studies. Biol Rev 73:43–78
Välimäki P, Kaitala A, Kokko H (2006) Temporal patterns in reproduction may explain variation in mating frequencies in the green-veined white butterfly *Pieris napi*. Behav Ecol Sociobiol 61:99–107
Välimäki P, Kivelä SM, Mäenpää MI (2011) Mating with a kin decreases female remating interval: a possible example of inbreeding avoidance. Behav Ecol Sociobiol 65:2037–2047
Vane-Wright D (2003) Butterflies. The Natural History Museum, London
Watanabe M (1979) Natural mortalities of the swallowtail butterfly, *Papilio xuthus* L., at patchy habitats along the flyways in a hilly region. Jpn J Ecol 29:85–93
Watanabe M (1981) Population dynamics of the swallowtail butterfly, *Papilio xuthus* L., in a deforested area. Res Popul Ecol 23:74–93
Watanabe M (1982) Leaf structure of *Zanthoxylum ailanthoides* Sieb. et Zucc. (Rutales: Rutaceae) affecting the mortality of a swallowtail butterfly, *Papilio xuthus* L. (Lepidoptera: Papilionidae). Appl Entomol Zool 17:151–159
Watanabe M (1983) Radial growth patterns of a pioneer tree, *Zanthoxylum ailanthoides* Sieb. et Zucc. (Rutales: Rutaceae), related to the population dynamics of a swallowtail butterfly, *Papilio xuthus* L. (Lepidoptera: Papilionidae). Jpn J Ecol 33:253–261
Watanabe M (1988) Multiple matings increase the fecundity of the yellow swallowtail butterfly, *Papilio xuthus* L., in summer generations. J Insect Behav 1:17–29
Watanabe M, Ando S (1993) Influence of mating frequency of lifetime fecundity in wild females of the small white *Pieris rapae* (Lepidoptera, Pieridae). Jpn J Entomol 61:691–696
Watanabe M, Ando S (1994) Egg load in wild females of the small white *Pieris rapae crucivora* (Lepidoptera, Pieridae) in relation to mating frequency. Jpn J Entomol 62:293–297
Watanabe M, Hirota M (1999) Effects of sucrose intake on spermatophore mass produced by male swallowtail butterfly *Papilio xuthus* L. Zool Sci 16:55–61
Watanabe M, Nozato K (1986) Fecundity of the yellow swallowtail butterflies, *Papilio xuthus* and *P. machaon hippocrates*, in a wild environment. Zool Sci 3:509–516
Watanabe M, Oh'ura T (1997) Egg cannibalism by newly hatched larvae of the small white butterfly, *Pieris rapae crucivora* (Pieridae), on an artificial diet. J Lepid Soc 51:304–315
Watanabe M, Omata K (1978) On the mortality factors of the lycaenid butterfly, *Artopoetes pryeri* M. (Lepidoptera, Lycaenidae). Jpn J Ecol 28:367–370
Watanabe M, Sato K (1993) A spermatophore structured in the bursa copulatrix of the small white *Pieris rapae* (Lepidoptera, Pieridae) during copulation and its sugar content. J Res Lepid 32:26–36
Watanabe M, Yamaguchi H (1993) Egg cannibalism and egg distribution of two *Pieris* butterflies, *Pieris rapae* and *P. melete* (Lepidoptera, Pieridae) on a host plant, *Rorippa indica* (Cruciferae). Jpn J Ecol 43:181–188. (in Japanese with English summary)
Watanabe M, Nozato K, Kiritani K (1986) Studies on ecology and behavior of Japanese black swallowtail butterflies (Lepidoptera: Papilionidae). 5. Fecundity in summer generations. Appl Entomol Zool 21:448–453
Watanabe M, Koizumi H, Suzuki N, Kiritani K (1988) Studies on ecology and behavior of Japanese black swallowtail butterflies. 7. Nectar of a glory tree, *Clerodendron trichotomum*, as a food resource of adults in summer. Ecol Res 3:175–180
Watanabe M, Bon'no M, Hachisuka A (2000) Eupyrene sperm migrates to spermatheca after apyrene sperm in the swallowtail butterfly, *Papilio xuthus* L. (Lepidoptera: Papilionidae). J Ethol 18:91–99
Wedell N, Cook PA (1998) Determinants of paternity in a butterfly. Proc R Soc Lond B 265:625–630

Wiklund C, Kaitala A, Lindfors V, Abenius J (1993) Polyandry and its effect on female reproductive output in the green-veined white butterfly (*Pieris napi* L.). Behav Ecol Sociobiol 33:25–33

Wiklund C, Kaitala A, Wedell N (1998) Decoupling of reproductive rates and parental expenditure in a polyandrous butterfly. Behav Ecol 9:20–25

Yamamoto M (1981) Comparison of population dynamics of 2 pierid butterflies, *Pieris rapae crucivora* and *Pieris napi nesis*, living in the same area and feeding on the same plant in Sapporo, northern Japan. J Fac Sci Hokkaido Univ Ser VI Zool 22:202–249

Zalucki MP, Kitching RL (1982) Dynamics of oviposition in *Danaus plexippus* (Insecta: Lepidoptera) in milkweed, *Asclepias* spp. J Zool (Lond) 198:103–116

Chapter 5
Mating Conflict

Keywords Copula duration • Desiccation • Last male precedence • Puddling • Resting period • Saline • Sphragis • Sodium ion • Spermatheca • Spermatophore

Although the nutrients for the reproductive output of females are amino acids from the host plants during the larval stage, sugars from nectar during the adult stage and ejaculates from the conspecific males during copulation, the only intake of amino acids and sugars in males is as major nutrients for somatic maintenance and ejaculate production. In addition, deriving saline from mud puddling behaviour could be an important role for male matings in some species, as well as for the muscular and nervous systems of the body itself. The most adequate host plants for nutrition might increase the male body in size; then, the large males are probably preferred by females when they are encountered because the large males transfer a larger amount of ejaculates, probably with a large amount of sperm. Sugars and saline are also important for males to produce an increased spermatophore mass, particularly mated males. Because the nutrients derived from the host plants could produce less than a half spermatophore at the subsequent mating, mated males have to resort exclusively to nectar during a certain period after mating to produce a full-sized spermatophore for the next mating. Because monogamous females can lay eggs throughout their lifespan, the quality of sperm until fertilisation has been somewhat of a concern. However, recent studies indicated that males can manipulate the quantity of eupyrene and apyrene sperm to transfer with his mating history, probably affected by the nutrition shortage in males. The large spermatophore transferred in the bursa copulatrix stimulates release of the female mate refusal posture against males showing courtship behaviour. The larger a spermatophore is transferred, the longer the unreceptive period of females might appear. In polyandrous species, therefore, males might evolve to produce a large spermatophore to avoid sperm competition among the sperm of different males. Thus, males of the European pierid butterflies transferred a single spermatophore that was about 25 % of body mass for the paternity.

M. Watanabe, *Sperm Competition in Butterflies*, Ecological Research Monographs,
DOI 10.1007/978-4-431-55945-0_5

Ejaculate Production of Virgin Males

Butterfly males offer females nuptial gifts, either to ensure effective mating or, in many species, as a paternal investment that increases reproductive success. Consequently, selection may favour males that produce relatively large nuptial gifts. Indeed, Svärd and Wiklund (1989) showed that the mass of the ejaculate is greater in more polyandrous butterfly species. However, the components of nuptial gifts are also typically limited in supply, which indicates that the amount of ejaculates provided to each female should be a key determinant of a male's reproductive success (Birkhead and Møller 1998). Therefore, mating must be costly for males, particularly if it increases their mortality or decreases their future reproductive success. Caballero-Mendieta and Cordero (2013) pointed out that mating costs can have ecological and physiological origins. The former includes disease transmission and predation risk, and the latter is the energetic costs of courtship or ejaculate production costs, which are especially important for males because they have to invest a substantial amount of resources. A negative effect of ejaculate production such as decrease in male longevity might be observed.

Males vary the size of spermatophores produced, depending on their mating history. It seems to be reasonable that males would invest more in the initial mating when the probability of future mating is low. Accordingly, a male's second spermatophore is usually significantly smaller than the one he produced the day before. Previously mated males of the swallowtail butterfly, *Papilio xuthus*, produce spermatophores that are about half the size of those they produced as virgins when they used nutrients derived only from their larval stages (Watanabe and Hirota 1999). Lederhouse et al. (1990) showed that subsequent spermatophores in the eastern tiger swallowtail, *Papilio glaucus* (Fig. 5.1), are only one third the volume of the first spermatophores. Therefore, internal reproductive organs of the mated males might be different in size, content or appearance from those of the virgin males, at least for a few days after the mating. In addition, rapid multiple

Fig. 5.1 Female eastern tiger swallowtail, *Papilio glaucus*, feeding on red clover

mating by males of the anise swallowtail, *Papilio zelicaon*, leads to smaller spermatophores deficient in sperm (Sims 1979). In the cabbage white, *Pieris rapae*, first ejaculates were larger than successive ones, which did not differ among themselves (Watanabe et al. 1998a). In addition, the cumulative mass of previous spermatophores was not correlated with that of the last mating. On the other hand, males of the monandrous speckled wood butterfly, *Pararge aegeria*, are willing to mate with another female soon after copulation, although the second spermatophore reaches only about a third of the mass of the first when the male mates twice on the same day (Vande Velde et al. 2011).

For many butterfly species, ejaculate size is correlated with male size (Bissoondath and Wiklund 1997). Even though at the first mating small virgin males transfer a small spermatophore, they might suffer reduced reproductive success for two reasons. The first is that females are more likely to re-mate when they receive a small ejaculate. The second is that males which transfer small ejaculates would suffer from reduced ability in sperm competition if the female subsequently re-mates with a relatively large male or virgin males.

A small spermatophore decreases the period of sexual unreceptivity in some lepidopteran females (Oberhauser 1989), during which time females lay eggs exclusively fertilised by the previous male. Then, the number of fertilised eggs by the male that transferred the small spermatophore decreased. In the alfalfa butterfly, *Colias eurytheme*, female fecundity has been shown to correlate positively with the size of the received ejaculate (Rutowski et al. 1987). In addition, small spermatophores might contain fewer eupyrene and apyrene spermatozoa than large ones (He and Miyata 1997). However, Proshold (1991) found that the older males of the tobacco budworm, *Heliothis virescens*, contained more eupyrene sperm bundles than younger males in the duplex the night after mating. The number of eupyrene sperm bundles in the spermatophore also did not differ between the first and the successive matings in the cabbage white, *Pieris rapae* (Watanabe et al. 1998a). Multiplying by 256, the male transfers about 11,000 eupyrene spermatozoa at every mating, whereas the first spermatophore contained about 46,000 apyrene spermatozoa, and the successive spermatophores contained higher numbers. Therefore, in the cabbage white, the sperm density increased after the first mating, although the spermatophore mass decreased.

Ejaculate Production of Mated Males

Swallow and Wilkinson (2002) reviewed that mated males transferred ejaculates that had higher proportions of eupyrene sperm but produced smaller ejaculates because they did not have time to recover their sperm volume. In cabbage white *Pieris rapae* males (Fig. 5.2), the spermatophore mass decreases with the number of matings, although the masses at second through fourth matings are not significantly different from each other (Watanabe et al. 1998a). Then, females that copulated initially with mated males might receive a smaller spermatophore and might

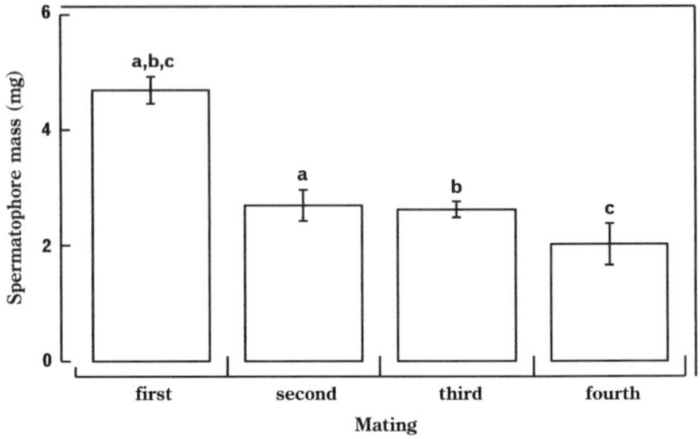

Fig. 5.2 Spermatophore mass transferred from males of the cabbage white, *Pieris rapae*, at first, second, third and fourth matings. *Bars* are standard errors. Same letters are significantly different at the 0.1 % level by Mann–Whitney *U* test. Number of matings exerts negative effect on spermatophore mass (Kendall test). (From Watanabe et al. 1998a)

re-mate sooner than females that mated initially with virgin males. Although constraints in ejaculate production rate are part of the cause, a decrease in ejaculate size could reflect a change in resource allocation strategies with age (Rönn et al. 2008). Reinhold et al. (2002) suggested that if young females are abundant in each discrete generation, males should transfer the larger ejaculate in early copulations, because female fecundity and the probability of achieving additional matings decreases with time and sperm competition intensity increases over time.

Summer generations of swallowtail butterflies are continuous because of the long flying season and their relatively long adult longevity (Watanabe and Kobayashi 2006). Consequently, there is a complex age structure of females with much variation in fecundity and mating status throughout the flying season. Therefore, males of polyandrous swallowtail species such as *Papilio xuthus* should allocate more equal amounts of ejaculate resources over successive matings, because they could encounter females of various mating status. However, the simplex of male butterflies just after copulation was always almost empty, suggesting that they cannot skimp on the amount of spermatophore materials transferred. Therefore, males of the swallowtail butterfly have improved their ability to produce spermatophore substances after the first mating, and they have to recover their spermatophore mass quickly before the next mating (Watanabe and Hirota 1999). For the speckled wood, *Pararge aegeria*, it takes up to 7 days after the first copulation before the male is able to produce a similarly sized spermatophore (Vande Velde et al. 2011). In addition, replenishment of the materials available to put into spermatophores can also take several days in monandrous species such as the pipevine swallowtail, *Battus philenor* (Rutowski et al. 1989).

Several changes in the characteristics of the male internal reproductive organs with mating experience have been found (Sasaki et al. 2015). Although in virgin males the simplex is long and contains white, granular material near the aedeagus and clear, pasty material elsewhere, the simplex after the first mating is shorter and of lower mass, and neither the white granular substance nor the clear pasty substance is found in the simplex. Their relative position in the simplex requires the white material to be placed first in the bursa copulatrix, followed by the clear substance. In swallowtail butterflies, spermatophore and accessory substances are transferred to the bursa copulatrix separately, and the accessory substance is located behind the spermatophore inside the bursa copulatrix (Watanabe et al. 2000). On the other hand, sperm is stored in the duplexes that unite to form the simplex (Hiroyoshi and Mitsuhashi 1999). Thus, sperm must be transferred to the female after transfer of the clear pasty substance that constructs the spermatophore.

As expected, males, especially mated males, of the polyandrous swallowtail butterfly, *Papilio xuthus*, produce a larger spermatophore than males of the monandrous species, *Papilio machaon*, suggesting that there is a trade-off relationship among production of ejaculate characters and that polyandrous species allocate more resources for production of the spermatophore. Bissoondath and Wiklund (1995) showed that the relative ejaculate mass increases with degree of polyandry across the butterfly species. Such a large ejaculate may have positive effects on female fecundity and negative effects on female sexual receptivity. When females received an smaller spermatophore, they tended to re-mate more often and sooner than average to maintain their own metabolism as well as to increase their fecundity (Watanabe 1988). Therefore, the sperm competition theory predicts that males will be selected to transfer larger spermatophores when females are polyandrous (Rutowski et al. 1983). It is already known that the risk of sperm competition incurred with rivalry varies because of variations in the size of the spermatophore transferred in respective copulations. Although males in polyandrous species produce large ejaculates to cope with the high degree of male competition in those species (Svärd and Wiklund 1989), the spermatophore size is not related to the number of eupyrene sperm in the cabbage white, *Pieris rapae* (Watanabe et al. 1998a).

Mated males produced spermatophores of lower mass with longer copula duration. For the cabbage white, *Pieris rapae*, the duration of first copulations was on average 59.8 min, whereas that for second copulations was 92.4 min, which is significantly longer than that of first copulation (Fig. 5.3). There are significant effects of mating numbers on the copula duration. Even though the copulating pair is not disturbed by any factors, the males tend to prolong copula duration with the numbers of mating. The copula duration of the fourth mating was twice that of the

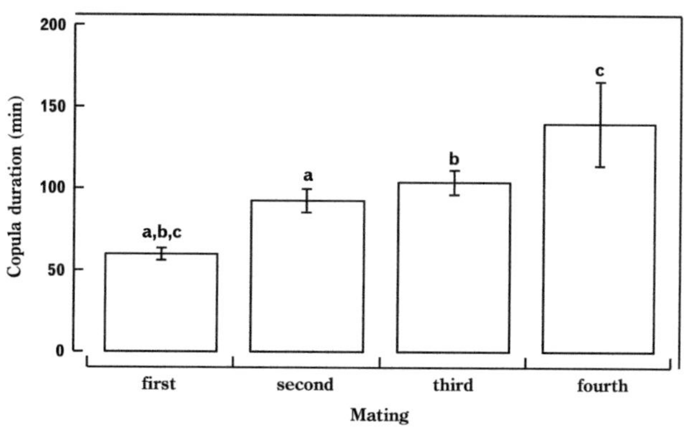

Fig. 5.3 Mean copula durations with males of the cabbage white, *Pieris rapae*, from the first to the fourth matings. *Bars* are standard errors. Same letters are significantly different at the 1 % level by Mann–Whitney *U* test. Number of matings exerts positive effect on copula duration (Kendall test). (From Watanabe et al. 1998a)

first matings, while such males seemed to maintain high sperm numbers to succeed in sperm competition (Watanabe et al. 1998a). Rutowski et al. (1987) showed that even the very small spermatophores transferred by recently mated males of the alfalfa butterfly, *Colias eurytheme*, can result in successful egg fertilisation and oviposition.

A mating might have a negative effect on male longevity, and this effect must be strong in males that suffer a resource restriction during the larval stage. In males of the lycaenid butterfly, *Callophrys xami*, that are resource limited by experimental food limitation of last-instar larvae, Cordero (2000) found that longevity of mated males was shorter than that of virgin males. Because sperm competition selects for large and nutritious ejaculates that delay female re-mating, males of polyandrous species must pay larger ejaculate production costs than males of monandrous species. On the other hand, multiple mating in the monarch butterfly, *Danaus plexippus*, has no effect on male longevity (Oberhauser 1989). There is no difference in longevity between the sexes of the monandrous nymphalid butterfly, *Lethe diana* (Takeuchi 2012). In the polyandrous pierid butterfly, *Pieris napi*, longevity decreases in multiple mated males, but no effects on ejaculate production are detected (Ferkau and Fischer 2006).

Nectar Feeding

The timing and source of nutritional resources allocated to reproduction have critical consequences for reproductive strategies of both males and females. Such allocated resources are derived from stored reserves during larval stages and current feeding in the adult stage. In the adult stage, many butterfly species are strict nectarivores. They need nectar as a vital day-to-day energy source, and visit flowers of a wide range of nectar-producing plant species; thus, they are opportunistic flower visitors that probe flowers of various sizes with open or concealed nectaries. They approach flowers with a loosely coiled proboscis and uncoil it after landing (Figs. 5.4, 5.5, and 5.6). The proboscis movement during flower handling might be stereotypic but enables it to adjust to various flower depths. The reason why adult butterflies feed on nectar is, at least, for water, energy, reproduction, defence, communication and neurotransmission (Vane-Wright 2003). Of six reasons, the main purposes are the first two. Generally, although butterfly adults have a small body and therefore have a relatively large surface area over which they lose water easily, the common habitats of most butterflies are often in full sun in warm areas, so that there is a potential for adults to be at risk of dehydration. Consequently, most butterflies require water.

Although the longevity of unfed females of the swallowtail butterfly, *Papilio xuthus*, is about 1 week with only 43 % of the fresh weight of 0-day-old ones (= just after eclosion), females taking water survived for 15 days (Watanabe 1992). Therefore, butterfly adults have to drink water, and nectar could often provide a sufficient supply as water. The second reason of nectaring is for obtaining energy,

Fig. 5.4 Female common straight swift, *Parnara guttata*, feeding on red clover

Fig. 5.5 Male common yellow swallowtail, *Papilio machaon*, feeding on wild lily. Pollen from the lily has adhered to the hindwings of the butterfly

because nectar is primarily a source of sugars, needed to power active flight, or for conversion to fat reserves. Butterfly adults generally find nectar by following the 'honey guides' in the colourful petals that point to the nectaries, sometimes using UV reflectance. With their uncoiled proboscis, they suck the nectar into their alimentary canal for quick digestion.

As shown in Fig. 5.7, females of the swallowtail butterfly, *Papilio xuthus*, fed on 10 % solution of sugar had a constant weight throughout their lifespan. Females fed on a 20 % solution of sugar become heavier as they age. Differences in sugar concentration affect daily sugar intake in each butterfly. If the amount of sugar was

Fig. 5.6 Female fritillary butterfly *Argyronome ruslana* feeding on red clover

effective for females to produce excess eggs compared to females given no sugar, the number of mature eggs would be related to the amount of accumulated sugar intake. However, almost no mature eggs are added in females feeding on solutions of sugar less than 1 %. During younger stages (Fig. 5.8), there is also no relationship between the number of mature eggs and the accumulated sugar intake. Egg maturation in younger females probably depends on the nutrition reserves carried over from the larvae rather than foods taken after emergence. The fat body seems to be a major energy resource for the maturation of eggs and for maintenance of adult life. However, because the fat body decreases as they age, older females would depend more heavily on sugar as an energy resource. Accordingly, it seems reasonable that the regression coefficients in the relationship between the number of mature eggs and the amount of accumulated sugar intake by females increase. Thus, nectar is the energy source for female butterflies, although it is low in protein content.

The female fat body is a nutrition reservoir derived from the larval stage and is a resource for egg maturation during the adult stage, suggesting that the number of mature eggs produced is determined by the amount of fat body at emergence. In the swallowtail butterfly, *Papilio xuthus*, there is a rapid egg maturation in females more than 2 days old feeding on sugar solutions of more than 10 % (Watanabe 1992). The aged females could allocate their available energy (fat body + sugars) for egg maturation and body maintenance.

Because different species of flowers must have different nectar compositions, nectar might be a slightly variable resource for the butterflies. In addition, within any species of flowering plants, the amount, concentration of sugars, and composition of nectar varies with each flower age, time of day, weather, and the activities of nectar-feeding insects. Consequently, the quality of nectar in any individual flower could be unpredictable for each butterfly until visiting. Three sugars—monosaccharide fructose, glucose, and disaccharide sucrose—are generally found in the nectar of many plant species, and the composition also varies not only among plant species but also with the age of a single flower, although many animals do not visit for nectaring (Watanabe et al. 1988). Fischer and Fiedler (2001) showed that

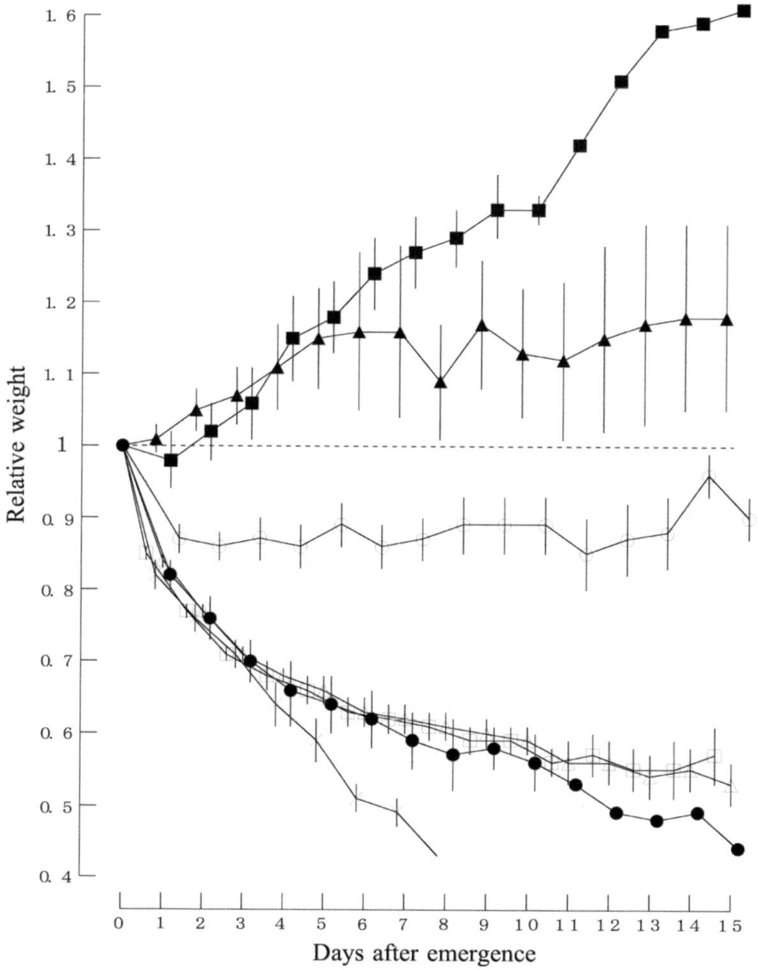

Fig. 5.7 Change in relative weight of females of the swallowtail butterfly, *Papilio xuthus*, for unfed (*cross*), feeding on water (*solid circle*), or on 0.1 % (*open square*), 1 % (*open triangle*), 10 % (*open circle*), 20 % (*solid triangle*) or 50 % (*solid square*) sugar solutions. (From Watanabe 1992)

the females of the purple-edged copper, *Lycaena hippothoe*, increased fecundity sevenfold when given sugar as opposed to only water.

Although the sugar concentration of the nectar in flowers partly depends upon weather conditions, it seems to be maintained within a wide range. Nectar is generally most dilute and abundant in the early morning, and by midafternoon it is most concentrated and scarce, a consequence of depletion by insect visitors and decreasing air humidity. Muraoka and Watanabe (1994) showed that the daily rhythms in sugar concentration of the flower of the field cress, *Rorippa indica* (Fig. 5.9), are independent of flower age but dependent upon nighttime dew and daytime evaporation (Fig. 5.10). Such wide changes in sugar concentration suggests

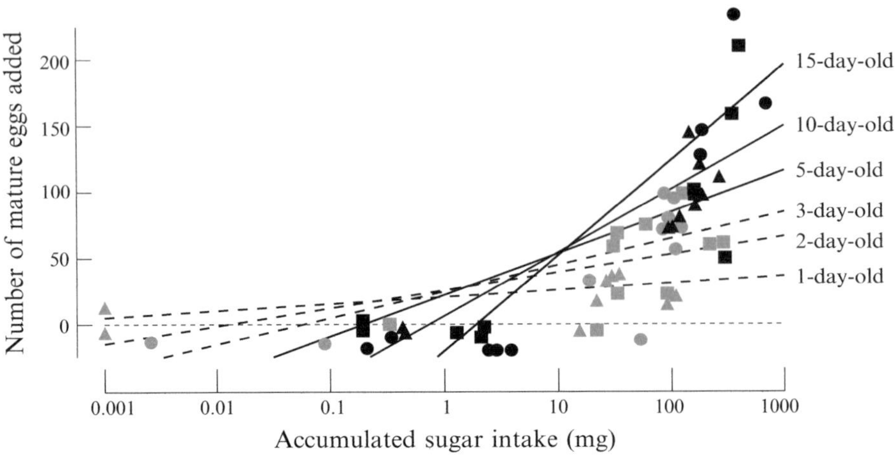

Fig. 5.8 Relationship between number of mature eggs added and accumulated sugar intake in female swallowtail butterfly, *Papilio xuthus*, after emergence. Regression coefficient is 5.1 for 1-day-old (*grey triangle*), 12.4 for 2-day-old (*grey square*), 20.5 for 3-day-old (*grey circle*), 29.9 for 5-day-old (*black triangle*), 49.1 for 10-day-old (*black square*) and 71.9 for 15-day-old (*black circle*) females, respectively. *Solid lines* and *broken lines* indicate significant and nonsignificant, respectively. (From Watanabe 1992)

that the nectar could be available to various species of insect visitors. For example, most pierid butterflies prefer a 20 % sugar solution, and they are active from 0600 to 1000, during which time the sugar concentration in the field cress was around 20 %. On the other hand, syrphid flies are observed to visit the field cress flowers to lick nectar of high sugar concentration in the afternoon. The Queen Alexandra's sulphur, *Colias alexandra*, prefers nectar with low sugar concentration, chiefly those including monosaccharides (Watt et al. 1974). The energy intake efficiency for some butterfly species is maximised at about 40 % sucrose solution (May 1985). Norris (1936) reported on the feeding regimes of the butterflies related to fecundity and survival depending upon the concentration level of sugar in nectar. Therefore, the nectar-feeding habit of the adult butterfly must have come to be evolved in the lifetime reproductive strategy.

Males of the swallowtail butterfly, *Papilio xuthus*, also feed on nectar for somatic maintenance as well as for increase in ejaculates (Watanabe and Hirota 1999). After 2 days of age, most virgin males fed only water ejaculate a spermatophore with weight similar to or lower than that of 1-day-old males (Fig. 5.11). However, at the respective ages, virgin males fed on sucrose solution produce significantly larger spermatophores than those fed water only. There is a positive relationship between the amount of sucrose ingested and the weight of the spermatophore ejaculated by virgin males. Although selection may favour the production of large spermatophores, the production of ejaculates is costly and limited. Therefore, in addition to mating history and body size of males, the time between

Fig. 5.9 Flowers of the yellow cress *Rorippa indica*

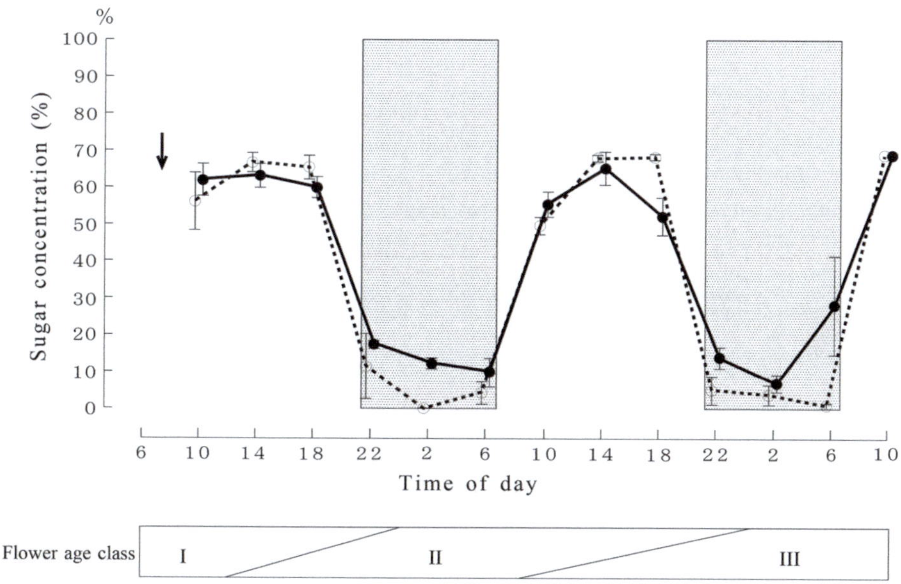

Fig. 5.10 Changes in sugar concentration in nectar in bagged (*closed circle*) and unenclosed (*open circle*) flowers of the field cress, *Rorippa indica*. *Arrow* indicates anthesis time; *shaded areas* represent periods of nighttime dew. (From Muraoka and Watanabe 1994)

previous matings (Sims 1979) and age (Oberhauser 1988) have been found to significantly affect spermatophore size.

Although the spermatophore size decreases in the second mating, the intake of sugar solution increases the weight of that spermatophore (Watanabe and Hirota 1999). Figure 5.12 shows that the second spermatophore produced by the mated males of the swallowtail butterfly, *Papilio xuthus*, fed on water is around 2 mg, which is significantly lighter than that of the first copulation, irrespective of their

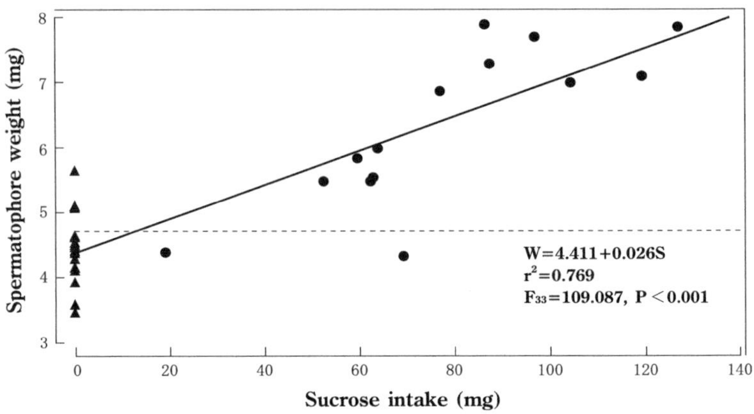

Fig. 5.11 Relationship between sucrose intake (*S*, mg) and spermatophore weight (*W*, mg) ejaculated by virgin male of the swallowtail butterfly, *Papilio xuthus*. *Dotted line* shows mean weight of spermatophore ejaculated by males on the day following emergence. *Circles* and *triangles* represent males feeding on 20 % sucrose solution and water, respectively. (From Watanabe and Hirota 1999)

resting period. Because the resting period does not increase the weight of the spermatophore, a male's ability to produce spermatophores during their lifetime without feeding on sucrose solution is apparently limited by the quantity of nutrients accumulated during the larval stage. On the other hand, the intake of sucrose solution increases the weight of the spermatophore at the second copulation when the resting time is more than 3 days between the first and second copulation. The resting period therefore increases the weight of the second spermatophore to about 6 mg.

Figure 5.13 shows the relationship between the amount of sucrose ingested and the weight of the second spermatophore relative to the first in the swallowtail butterfly, *Papilio xuthus*. When males are fed only water, the second spermatophore is about half the size of the first. However, feeding on sucrose increases the second spermatophore to as much as 1.5 times that of the first spermatophore. The regression line indicates that ingesting 77.57 mg sucrose (or 390 mg 20 % sucrose solution) allows a male to produce a second spermatophore that is of similar weight to the first. Given that males consume about 90 mg solution daily, they need about 4.3 days to produce a second spermatophore of the same weight as the first.

Before production of a second spermatophore of full size in the swallowtail butterfly, *Papilio xuthus*, a resting period of several days is required, during which the male's activity must consist of such activities as feeding, flying and avoiding predators. When the male re-mates, he has to produce another spermatophore of sufficient size with a sufficient number of sperm to fertilise as many eggs as possible. Sasaki and Watanabe (2015) reported that the number of resting days after copulation significantly affected the number of both eupyrene sperm bundles and apyrene sperm produced. For the former, 5 days after the first copulation, the mated male accumulated approximately 150 eupyrene sperm bundles, whereas males mated 1 day after the first copulation produced 34 bundles. For the latter,

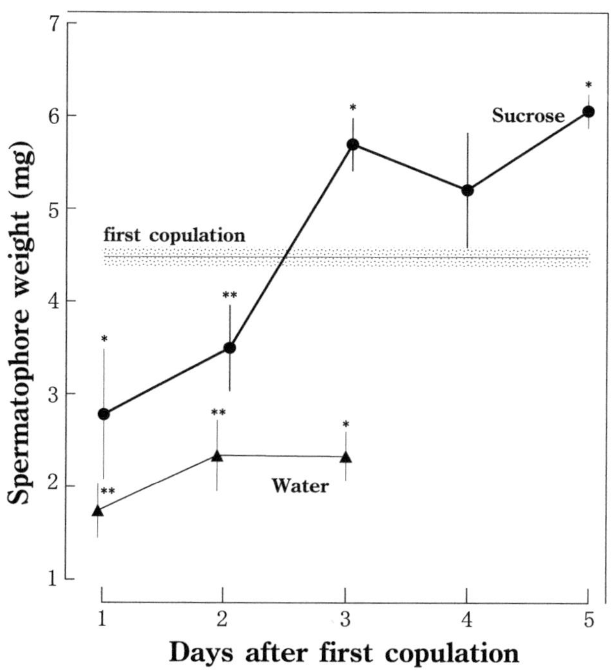

Fig. 5.12 Weight of spermatophore (mg, ±SE) produced at second copulation in relationship to time elapsed since first copulation in the swallowtail butterfly, *Papilio xuthus*. *Shaded area* including a *thin line* shows spermatophore weight ejaculated at first copulation by virgin males on day after emergence (with SD). *Circles* and *triangles* represent males fed on sucrose solution and on water only, respectively, during the resting period, which is indicated as days following first copulation (that is, 0 days means that a male copulated on 2 successive days). $*P < 0.05$ and $**P < 0.01$ in comparison to spermatophore weight at first copulation (Mann–Whitney *U* test). (From Watanabe and Hirota 1999)

5 days after initial copulation, the mated males accumulated approximately 500,000 apyrene spermatozoa, compared to 163,000 for males re-mated 1 day after the first copulation. Therefore, the duration of the male's recovery time and the rate of recovery in terms of spermatophore size and sperm number are important for males under sperm competition. The rapid increase in the spermatophore size after copulation must be adaptive for frequent matings of males in the polyandrous mating system.

Beyond Nectar

For some butterfly species, rotting fruits can supply not only sugars but also alcohols for energy, both of which might be good resources to help with reproduction in males and females. Carrion, mammal dung and bird droppings are also attractive for feeding, particularly by nymphalid butterflies, probably to supplement

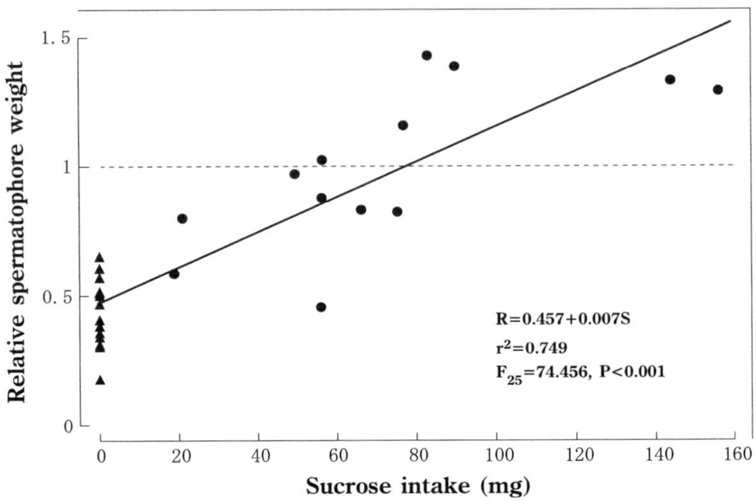

Fig. 5.13 Relationship between sucrose intake (S) during resting period and relative weight of spermatophore (R) ejaculated at the second copulation in the swallowtail butterfly, *Papilio xuthus*. *Dotted line* is that expected if weight of the spermatophore is the same at the first and second copulations. *Circles* and *triangles* represent males feeding on sucrose solution and water, respectively, during resting period. (From Watanabe and Hirota 1999)

nitrogen uptake (Beck et al. 1999) as well as sodium intake (Arms et al. 1974), and another important component of egg production (Adler and Pearson 1982). Lewis and Wedell (2007) found that males of the squinting bush brown, *Bicyclus anynana*, fed on fruit produced larger spermatophores on their first mating compared to males fed sugar only. Those fruit nutrients in the spermatophore seem to be passed on to females for use in egg production. However, fruit-fed males live shorter lives than the males fed on sugar only, suggesting that there is a trade-off between reproduction and somatic maintenance because spermatophore production is costly. Fischer et al. (2004) clarified that fruit serves as a carbon source for egg production in a similar manner as nectar. On the other hand, Vane-Wright (2003) stated that milkweed butterflies are strongly attracted to sources of pyrrolizidine alkaloids, the poisonous chemicals that make ragwort so dangerous to horses. Those alkaloids are released in the nectar, but they are more concentrated in the leaves and roots; then adult butterflies repeatedly probe the roots, probably to use for defence and making sex pheromones.

Butterfly species, particularly in *Papilio* spp., are a common group that demonstrate male puddling behaviour (Fig. 5.14). In Japan and Southeast Asia, males of black swallowtail butterflies often flock on damp sand or mud, being active throughout their adult lives in terms of puddling behaviour. Some of these butterflies visit puddles singly. Butterflies that land on the mud suck the tainted water of the mud. It is often observed that as fast as the muddy water is sucked up by the proboscis, jets of fluid are ejected from the anus. Vane-Wright (2003) described that in some species the proboscis is repeatedly extended backwards between the legs,

Fig. 5.14 Fresh males of the alpine black swallowtail, *Papilio maackii*, puddling

Fig. 5.15 Laboratory experiment on artificial feeding of adult swallowtail butterfly. Known volume of water, sugar solution or saline solution is on the coiled proboscis. Amount of ejection by the butterfly can be collected by filter paper under anus

and some of the ejected fluid is sucked up and thus 'recycled.' Thermal environmental conditions have been shown to influence butterfly puddling (Rawlins 1980); for example, hot weather causes butterflies to drink water along streams (Beck et al. 1999).

Three goals of puddling behaviour have been proposed by a number of researchers since Norris (1936). The first is to access a water supply, although these butterflies also show a 'pumping behaviour' to discharge water, ejecting from the anus (Fig. 5.15). Mazer and Appel (2001) reported that the water content of females was less than that of males, possibly accounting for the fact that only males puddle. The second goal of puddling behaviour is to achieve a reduction in body temperature. In fact, puddling has commonly been observed in tropical regions and in the hot summer season of temperate regions (Beck et al. 1999). However, very few females visit puddles, in spite of high ambient temperature (Frey et al. 2002), and extremely male-biased sex ratios have been described in reports on puddling activity (Adler 1982). The third putative goal of puddling is the intake of certain substances dissolved in the water. From this perspective, puddling is often viewed as differential foraging that supplies either ionic or nutrient resources needed by males of the eastern tiger swallowtail, *Papilio glaucus* (Arms et al. 1974). Adler and Pearson (1982) also pointed out that males of the cabbage white, *Pieris rapae*, preferred a solution including sodium ions as well as nectar. In general, insects must have sodium for their nerves to work effectively. However, the individuals that puddle are exclusively young males; older males do not visit puddles.

Downes (1973) demonstrated that additional sodium ions are required for the flight muscles of males to become active. Adler and Pearson (1982) showed that sodium ions were transferred from males to females during copulation and were exploited as nutrients for eggs. The hatching rate of the eggs deposited by females that copulated with males that had puddled increased (Pivnick and McNeil 1987). This result suggests one reason why females do not puddle; namely, they can acquire enough ions from their mate during copulation. Nevertheless, this is not an explanation for the observation that only young males visit puddles.

The neotropical butterflies, *Heliconius* spp., rely on pollen, in addition to floral nectar, as a food source (Krenn 2010). Adults gather a ball of pollen and hold it within the lightly coiled proboscis. The pollen ball is agitated in a fluid, and the resulting liquid is ingested by the adults. Because some amino acids are extracted from the pollen grains, adults could obtain the nitrogen for somatic maintenance and egg production. Consequently, pollen feeding completely alters the life history in comparison with the related butterfly species that are solely nectivores: oogenesis is continuous throughout the life, rather than stopping before adult emergence, and lifespan is increased (Dunlap-Pianka et al. 1977).

Males of many non-pollen-feeding butterfly species do not obtain protein available for spermatophore production. Certain tropical butterflies even appear to supplement their nitrogen budget by visiting protein-rich mud puddles (Beck et al. 1999). Although the potential importance of male-derived nutrients is expected to correlate positively with the mating success of males, sodium ions are not a nutrient. Therefore, the aim of sodium ion intake is assumed to be increased mating-related activities, such as searching behaviour for females, the volume of ejaculation, the number of spermatozoa produced, and other related factors.

Watanabe and Hirota (1999) demonstrated that the daily intake of distilled water of the males of the swallowtail butterfly, *Papilio xuthus*, was approximately 45 mg for 5 days after emergence, with decrease in body weight, and that the reproductive potential of males was limited unless their diet was supplemented with sucrose

Table 5.1 Copula duration and ejaculates of 3-day-old males of the swallowtail butterfly, *Papilio xuthus*, feeding on different concentrations of saline solution (±SE)

Saline solution	Copula duration (min)	Spermatophore mass (mg)	Accessory substances mass (mg)	Number of eupyrene sperm bundles	Number of apyrene spermatozoa (×10,000)
Distilled water	58.7 ± 2.2 (12)	5.6 ± 0.1 (11)	5.9 ± 0.3 (7)	66.9 ± 10.5 (9)	38.6 ± 4.8 (6)
0.001 mol	59.6 ± 6.8 (13)	6.0 ± 0.3 (13)	6.5 ± 0.7 (13)	82.2 ± 11.1 (13)	38.1 ± 5.9 (7)
0.01 mol	62.3 ± 3.4 (12)	6.5 ± 0.2** (13)	5.9 ± 0.4 (12)	70.7 ± 10.0 (13)	40.2 ± 5.1 (5)
0.1 mol	61.3 ± 5.7 (13)	5.7 ± 0.6 (13)	5.9 ± 0.6 (13)	69.8 ± 10.1 (13)	45.3 ± 7.8 (6)
1 mol	77.1 ± 10.4* (11)	5.3 ± 0.5* (11)	6.0 ± 0.7 (11)	65.5 ± 10.0 (11)	32.8 ± 5.2 (6)

*Significant difference from control males feeding on distilled water using Mann–Whitney *U* test at probability of $0.05 > P > 0.01$

**Significant difference from control males feeding on distilled water using Mann–Whitney *U* test at probability of $P < 0.01$

Number in parentheses is number of males examined

Source: Watanabe and Kamikubo (2005)

during adulthood. Consequently, males frequently visit nectar sources that also contain a small quantity of lipids and amino acids, but no saline (Baker and Baker 1973). As well as nectar sources, however, the swallowtail butterfly males land on mud puddles, which may provide some amount of saline. Watanabe and Kamikubo (2005) showed that 0.01 mol saline solution appeared to have been the most effective concentration for males on ejaculate production. Sodium must enhance the production of the spermatophore (Table 5.1). Arms et al. (1974) reported that sodium ions were a stimulus for the puddling behaviour of the males of the eastern tiger swallowtail, *Papilio glaucus*, which aggregated on moist soil containing 0.1 mol saline.

Adler and Pearson (1982) showed that the males of the cabbage white, *Pieris rapae*, may experience a net salt loss within several days after emergence, probably caused by spermatophore production. Although a considerable amount of saline solution is discharged during the puddling behaviour, mated males of the swallowtail butterfly, *Papilio xuthus*, intake salt as well as sugar (Table 5.2 (Niihara and Watanabe 2009). Saline intake also recovers the ejaculate mass of the mated males. Because a large spermatophore and a large number of sperm must be advantageous to the male under sperm competition in female polyandry, the puddling behaviour might be important to increase the reproductive success of males.

Pivnick and McNeil (1987) showed that males of the European skipper, *Thymelicus lineola*, stored enough sodium derived from larval feeding to mate successfully even without adult supplementation, whereas in many butterfly species, both sexes generally feed on more protein-rich food in the larval stage than in the adult stage, although the composition of lepidopteran spermatophores is approximately 85 % water and only approximately 7 % protein (Boggs 1981). To

Table 5.2 Quantity of intake and discharge for distilled water (DW), 20 % sucrose solution or 0.01 M saline solution by 1-day-old males of the swallowtail butterfly, *Papilio xuthus*, after first copulation (mg, ±SE)

	n	Intake	Discharge
DW	30	39.2 ± 6.98	12.8 ± 5.40
20 % sucrose solution	31	147.8 ± 6.90	0.1 ± 0.10
0.01 M saline solution	29	90.0 ± 20.53	63.2 ± 20.26
0.01 M saline solution	14	111.9 ± 34.83	82.1 ± 34.33
+20 % sucrose solution[a]	14	139.4 ± 13.13	0.0 ± 0.00

n number of males examined
[a]Males fed on 20 % sucrose solution after 0.01 M saline solution
Source: Niihara and Watanabe (2009)

transfer a spermatophore with accessory substances to the female's bursa copulatrix, males have to use resources in the fat body derived from larval stages.

Transfer and Storage of Sperm

The bursa copulatrix can be expanded to accommodate larger spermatophores, and the spermatophore during copulation is initially gel like (Watanabe and Sato 1993), suggesting there is little limit to the number of sperm that can be transferred into the spermatophore. Because the transfer of large numbers of sperm to the female at mating is advantageous in sperm competition, males might try to increase the spermatophore size as well as the number of sperm, even when sperm are going to mix in the spermatheca of the female. For males, in fact, transferring a large ejaculates provide several advantages in polyandrous species (McNamara et al. 2009). First, the contents of the ejaculates, that include sperm and seminal proteins as the nutrients for female life must increase female reproductive output (Boggs and Gilbert 1979), resulting in direct increase of male reproductive success. Second, the transfer of such a large spermatophore may reduce the likelihood of sperm competition because there is a positive correlation between ejaculate size and the time between subsequent matings for females (Oberhauser 1997). Finally, in the presence of sperm competition, an increased ejaculate size with high sperm number must increase the probability of fertilisation, resulting in high proportional paternity success (Simmons 2001). If the accelerated rate of production of spermatophore materials after mating in the polyandrous swallowtail butterfly, *Papilio xuthus*, is an adaptation that evolved in response to potentially brief intermating intervals (Watanabe and Hirota 1999), males should also accelerate the rate of production of sperm after copulation.

Compared to the size plasticity of the spermatophore, the spermatheca has a fixed storage capacity (Watanabe and Hachisuka 2005). Suzuki et al. (1996) found that the spermatheca was filled by sperm after a single mating in the silkworm, *Bombyx mori*. Although many sperm disappear during their migration from spermatophore to spermatheca (Watanabe et al. 2000), the number of sperm that safely arrive at the spermatheca should increase with the number of sperm transferred. When the spermatheca becomes filled before the migration ends, additional sperm must go to waste. If so, to transfer more sperm than can be contained in the spermatheca is maladaptive for males. On the other hand, because eupyrene sperm in the duplex of the male may retain high fertility for a long time (Hiroyoshi 1995), saving sperm by the males might be beneficial, even though the probability of future additional matings is low.

Eupyrene sperm bundles are released from the testis and are transported, via the vas deferens, to the duplex where they are stored until mating, and the sperm release follows a daily rhythmic pattern, resulting in an increased number of sperm present in the duplex as the number of days until mating increases (LaChance et al. 1977). Wedell and Cook (1999) pointed out that males are not dependent on the number of sperm released from the testis on the day of copulation. Males of some species have been shown to not ejaculate all the sperm that are present in the duplex in a single mating (Hiroyoshi 1995). In the swallowtail butterfly, *Papilio xuthus*, the sperm transfer ratio was low in virgin males that produced more sperm, although the number of sperm transferred increased with the number of sperm produced (Sasaki and Watanabe 2015). Virgin males might regulate the ratio of sperm transferred in relationship to their own sperm reserve for future matings.

Because of the lower probability of expected future matings, older males are apt to cease to discriminate between females in some insect species (Engqvist and Sauer 2001). In addition, for polyandrous species, virgin males must be unlikely to encounter virgin females as well as females mated to previously mated males, although both virgin and mated females have a high likelihood of accepting males. Therefore, the change in the rate of sperm transferred with mating by the mated males of the swallowtail butterfly, *Papilio xuthus*, might be a response to the low probability of expected future mating (Sasaki and Watanabe 2015). Saving sperm for subsequent mating may be also maladaptive for them when the risk of sperm competition is high. Wedell and Cook (1998) clarified that the males of the cabbage white, *Pieris rapae*, produced significantly more sperm on their second mating than males mating for the first time, although they transferred a smaller spermatophore. Although there is a clear effect of mating order on paternity (last male precedence), mated males producing more sperm might be better in competition with virgin males producing larger spermatophores but with fewer sperm. Therefore, increasing sperm numbers must be rather advantageous in sperm competition.

Females could benefit by mating with a virgin male if they need to receive sufficient nutrients incorporated into both eggs and female somatic tissue. In species in which the male ejaculate is used for fertilisation of eggs only, females may benefit from mating a second time if the sperm received in the first mating is not sufficient to fertilise all the eggs (Lederhouse 1981). However, males of the cabbage white, *Pieris rapae*, seem to reduce their nutrient investment to the female, maintaining high sperm numbers to succeed in sperm competition (Watanabe et al. 1998a). Rutowski et al. (1987) showed that even small spermatophores transferred by recently mated males of the alfalfa butterfly, *Colias eurytheme*, result in egg fertilisation and oviposition. Thus, sperm number is an important determinant in achieving fertilisation success in sperm competition (Parker 1970). However, as a rule, one spermatophore contains enough sperm to fertilise all the eggs of a female butterfly (Svärd and Wiklund 1989). In the cabbage white, Watanabe and Ando (1993) estimated the lifetime output of the female as about 250 eggs, whereas at least 10,000 eupyrene sperm were contained in a spermatophore at each mating (Watanabe et al. 1998a). Males may always transfer an excess number of eupyrene sperm to inseminate all a female's eggs at one mating. Therefore, certain mechanisms of sperm competition may also select for sperm velocity and power (Gage 1994). Sperm survival in the spermatheca of the female may be an additional important factor in sperm competition. In the large white, *Pieris brassicae*, only inactive sperm is found in the spermatheca 10 days after mating (Tschudi-Rein and Benz 1990).

Watanabe and Hachisuka (2005) reported that approximately 800 to 1000 eupyrene spermatozoa arrived in the spermatheca 1 day after the copulation of either ovipositing or inhibited females of the swallowtail butterfly, *Papilio xuthus* (Fig. 5.16). Their numbers then gradually decrease in both females, allowing or inhibiting oviposition. There are usually fewer eupyrene spermatozoa in the spermatheca of females allowing oviposition than in that of those inhibiting females. The spermatheca of ovipositing females 6 days after copulation has a lower number of eupyrene spermatozoa than does that of inhibiting females. At 7 days after copulation, no eupyrene spermatozoa are found in the spermatheca of ovipositing females, whereas about 300 eupyrene spermatozoa remain in that of inhibiting females. In addition, although apyrene spermatozoa in the spermatophore outnumber eupyrene spermatozoa 29 to 1, about 1100 apyrene spermatozoa are found in the spermatheca of ovipositing females, the number of which is roughly equal to that of eupyrene spermatozoa. Although about 600 apyrene spermatozoa are found in the spermatheca of inhibiting females, their number in both ovipositing and inhibiting females also decreases gradually as in the case of the eupyrene spermatozoa. However, a certain small number of apyrene spermatozoa (approximately 100) remains in both types of females.

Paternity Guards

Depending upon the species, males may respond to the threat of sperm competition by attempting to limit the access of other males to their mates by methods such as mate guarding and by secreting materials, typically called mating plugs, that block the female's copulatory opening. Because females do not oviposit just after copulation, and because the host plants on which they oviposit are not always near the copulation site, mate guarding behaviour by males has little effect on the females. Although the copulation position inhibits rival males from mating because the male himself is a mating plug, the females cannot oviposit or perform active flight, resulting in increase of predation risk.

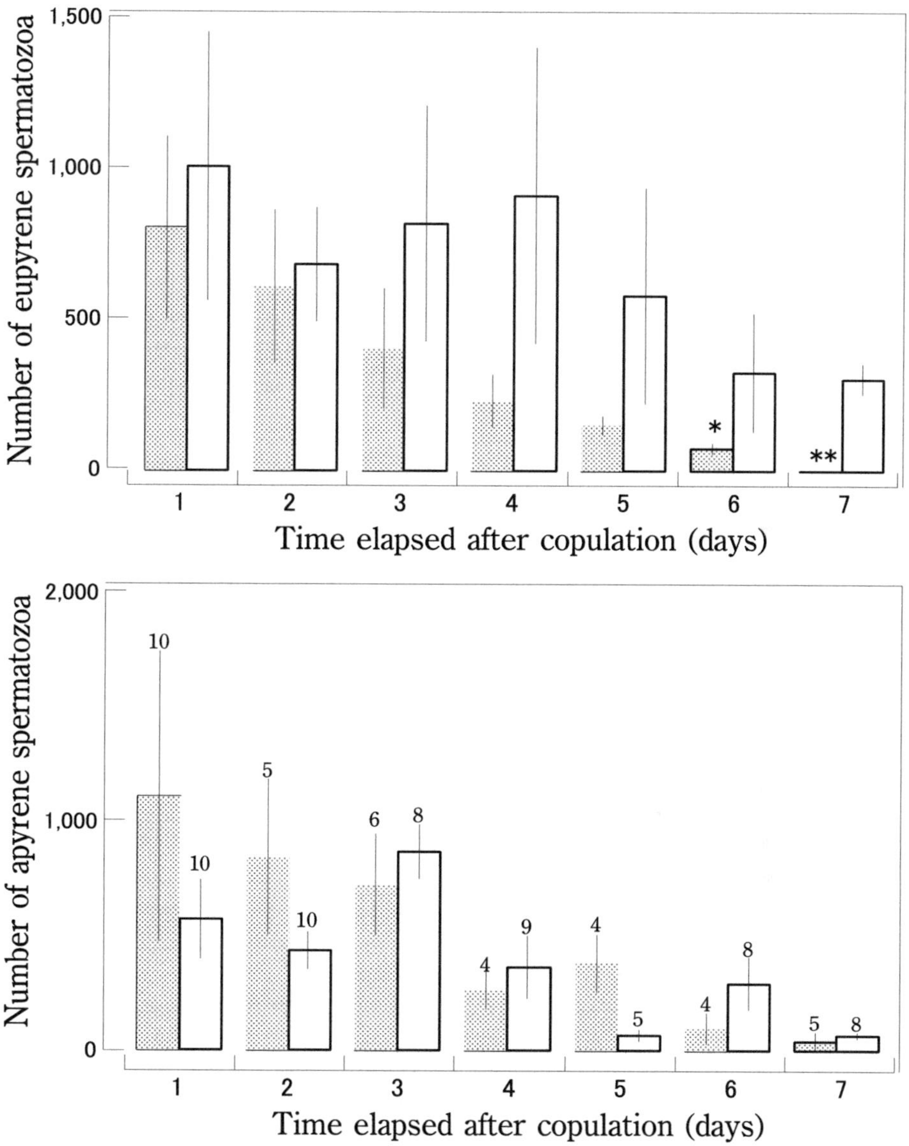

Fig. 5.16 Changes in number of eupyrene and apyrene spermatozoa in the spermatheca of female swallowtail butterfly, *Papilio xuthus*, allowing oviposition (*filled*) after copulation, in comparison with that of females inhibiting oviposition (*white*). *Bars* are SE. *Numbers above the bars* show sample size. (From Watanabe and Hachisuka 2005)

For males of some species, as the copulation is completed, the male plugs the female's vestibulum, using an additional accessory gland secretion that spills out of the female's copulatory opening to form a mating plug. This secretion hardens within a few hours after the copulation. Because female butterflies have two genital

Fig. 5.17 Mating plug placed on tip of female abdomen in the Japanese parnassian, *Parnassius citrinarius*

openings, males can block the copulatory opening without hindering oviposition. Therefore, no other male can gain access, but females can oviposit.

The mating plug seems to function as a means of reduced likelihood of female re-mating. The plug is placed on the mating pore of the female at copulation to physically prevent the female of several monandrous butterflies, such as the Chinese windmill, *Byasa alcinous*, and the Japanese apollo butterflies, *Parnassius* spp. (Matsumoto and Suzuki 1995), from re-mating: this plug is the so-called sphragis (Fig. 5.17). Accordingly, many sperm from the first mated male will be used for insemination by the female. The sphragis might also function as a visual deterrent to rival males, as the larger the plug the less likely the rival males will attempt to mate with the females (Dickinson and Rutowski 1989). However, it is not completely effective as females are capable of re-mating despite the presence of a plug. Vane-Wright (2003) pointed out that this plug may eventually dissolve or fall out, so that long-lived females of these species can benefit from a second mating. Thus, reducing the female receptivity for re-mating by means of apyrene sperm in the spermatheca must be rather important even in monandrous butterflies in which males produce mating plugs.

Wedell (2005) showed another way in which males can reduce female receptivity by rendering them unattractive to rival males following the mating. In the queen butterfly, *Danaus gilippus*, males release a pheromone-bearing dust from their hair-pencils containing a flight inhibitor and a glue that sticks the inhibitor onto the female's antennae, reducing their attractiveness. Consequently, the pheromones reduce the likelihood that the female will mate again, and also reduce costly harassment by additional males, which is beneficial to the females.

Males of most butterfly species apparently lack means to decrease competing sperm physically from the females, because they leave only a small sperm plug after copulation. Furthermore, males have no provision for removing a rival male's sperm. Their penis cannot reach the spermatheca, where eupyrene sperm is stored. The size and the shape of the spermatophore in the bursa copulatrix induce female mate refusal posture using physical stimulation. Females that receive a larger spermatophore subsequently exhibit longer refractory periods, and this extends the oviposition period in which the offspring are sired exclusively by the first male's sperm because sperm competition is avoided. The physical presence of the spermatophore in the bursa copulatrix is sufficient to switch off female receptivity. Thus, males may evolve to increase spermatophore mass, particularly for the species with short longevity of adults or with a large proportion of eggs laid in the days following mating (McNamara et al. 2009). Vahed (1998) pointed out that the transfer of a large ejaculate by the male can counter the effects of sperm competition directly by diluting or displacing rival sperm, thereby increasing the proportion of the female's eggs fertilised by the male, or indirectly by extending the duration of the female's post-mating refractory period, or by increasing the rate of oviposition following mating. However, because females absorb spermatophore nutrients, the spermatophore size begins gradually to decrease about 2 days after copulation.

Because a single mating provides sufficient sperm for a female's lifetime egg production (Rutowski 1984), male re-mating frequency is likely to be an index of the risk of sperm competition affecting male investment in spermatogenesis, as well as nutrients. With increased risk of sperm competition, to ensure optimum fertilisation males should invest more in their ejaculate. After mating, eupyrene sperm break from their bundles and migrate together with, or after, apyrene sperm from the spermatophore to the spermatheca. LaChance et al. (1977) reported that the number of sperm bundles accumulated in the duplex in the male increased with age before mating.

Several potential costs that males incur in production of spermatophores have been suggested, such as the time cost of mating itself and the energy cost to replenish accessory gland materials to form the spermatophore (Dewsbury 1982). Spermatophores contain significant quantities of nitrogen (Marshall 1985) and sugars (Watanabe and Sato 1993). In addition, there is an effect of spermatophore size on delaying female re-mating (Sugawara 1979). Delayed female re-mating is beneficial to the first male if some proportion of eggs is fertilised by sperm of the second male. Large spermatophores delay female re-mating (Oberhauser 1989).

Although larger males produced a larger spermatophore on their first mating, multiple mating by males leads to smaller spermatophores but a higher number of sperm than that produced by the virgin males (Cook and Wedell 1996). In some butterfly species, the accessory substances contain nutrients that are incorporated into the female's eggs before fertilisation (Boggs and Gilbert 1979), resulting in increased female fecundity (Watanabe 1988), and may act as paternal investment. Kaitala and Wiklund (1994) showed that females receiving large spermatophores have a longer period of sexual unreceptivity. Therefore, a large spermatophore may be of advantage in sperm competition.

The interval of matings and the morphology of fertilisation ducts allow eupyrene spermatozoa from the second mating to be positioned close to the exit because of the cul-de-sac spermatheca in the swallowtail butterfly, *Papilio xuthus* (Watanabe et al. 2000). Last-male priority has been established as the predominant pattern of sperm use in most butterfly species that have been studied (Sims 1979).

Females of the monandrous swallowtail butterfly, *Papilio machaon*, seem not to use the spermatophore and accessory substances for either energy or egg production (Watanabe and Kobayashi 2006). They have larger fat bodies in their abdomens after eclosion than the polyandrous swallowtail butterfly, *Papilo xuthus*. This finding corresponds with body weights of the males. Males of *P. machaon* have clearly more fat body than male *P. xuthus*, even though their body sizes (= length of forewings) are similar. Therefore, non-virgin females of *P. machaon* must have required no further ejaculate from males, and thus would avoid their conspecifics or refuse mating when males are encountered. Because only virgin females permit copulation, the operational sex ratio must be biased towards males. The probability of mating in males may be low in the field, but the probability of re-mating by mated males must be zero if females are able to detect mated males. Then, males of *P. machaon*, particularly virgins, might increase the number of eupyrene spermatozoa as well as the maximum ejaculation mass as much as possible when they copulate. Consequently, sperm size would decrease.

For a male butterfly second mating, in general, the probability of encountering non-virgin females who can easily accept copulation must remain high in nature because of the long flying season and continuous generations during the summer. Hence, for the polyandrous swallowtail butterfly, *Papilio xuthus*, it may be advantageous to provide these females with large numbers of eupyrene spermatozoa in spite of reduced nutrient donations. The number of eupyrene sperm bundles in the spermatophore significantly increases for the second mating. During sperm competition, an increased risk with the degree of polyandry and production of high numbers of sperm may be advantageous. In the cabbage white, *Pieris rapae*, males transfer significantly more eupyrene sperm at their second mating, although the spermatophore size is reduced (Cook and Wedell 1996). Because there might be little sperm mixing in *P. xuthus* as a consequence of the apyrene sperm in spermatheca (Watanabe et al. 2000), it appears advantageous for males to transfer an excess of eupyrene spermatozoa with an excess number of apyrene spermatozoa at the second mating.

Kaitala and Wiklund (1995) showed that copulations with recently mated males of the green-veined white, *Pieris napi*, can last up to 24 h, suggesting that repeated matings may carry substantial time costs. In the alfalfa butterfly, *Colias eurytheme*, variation in the duration of copulations observed in the field suggests that males mate about once every 5 days (Rutowski and Gilchrist 1986). It is important for a male to be able to re-mate as soon as possible after copulation to produce a spermatophore of sufficient size to maximise his reproductive success; this is necessary because the genetic material of the female's last partner tends to predominate in the offspring (Simmons and Siva-Jothy 1998). Cook and Wedell (1996) demonstrated that the size of the spermatophore and the number of sperm depend on the male's mating history. With increased risk of sperm competition in polyandrous females, to ensure optimum fertilisation, eupyrene sperm must break from their bundles (Watanabe et al. 1998b) and migrate from the spermatophore to the spermatheca as soon as possible.

References

Adler PH (1982) Soil- and puddle-visiting habits of moths. J Lepid Soc 36:161–173

Adler PH, Pearson DL (1982) Why do male butterflies visit mud puddles? Can J Zool 60:322–325

Arms K, Feeny P, Lederhouse RC (1974) Sodium: stimulus for puddling behavior by tiger swallowtail butterflies, *Papilio glaucus*. Science 185:372–374

Baker HG, Baker I (1973) Amino acids in nectar and their evolutionary significance. Nature (Lond) 241:543–545

Beck J, Muhlenberg E, Fielder K (1999) Mud-puddling behavior in tropical butterflies: in search for proteins or minerals? Oecologia (Berl) 119:140–148

Birkhead TR, Møller AP (1998) Sperm competition and sexual selection. Academic, London

Bissoondath CJ, Wiklund C (1995) Protein content of spermatophores in relation to monandry/ polyandry in butterflies. Behav Ecol Sociobiol 37:365–371

Bissoondath CJ, Wiklund C (1997) Effect of male body size on sperm precedence in the polyandrous butterfly *Pieris napi* L. (Lepidoptera: Pieridae). Behav Ecol 8:518–523

Boggs CL (1981) Selection pressures affecting male nutrient investment at mating in heliconine butterflies. Evolution 35:931–940

Boggs CL, Gilbert LE (1979) Male contribution to egg production in butterflies: evidence for transfer of nutrients at mating. Science 206:83–84

Caballero-Mendieta N, Cordero C (2013) Male mating costs in a butterfly that produces small ejaculates. Physiol Entomol 38:318–325

Cook PA, Wedell N (1996) Ejaculate dynamics in butterflies: a strategy for maximizing fertilization success? Proc R Soc Lond B 263:1047–1051

Cordero C (2000) Trade-off between fitness components in males of the polygynous butterfly *Callophrys xami* (Lycaenidae): the effect of multiple mating on longevity. Behav Ecol Sociobiol 48:458–462

Dewsbury DA (1982) Ejaculate cost and male choice. Am Nat 119:601–610

Dickinson JL, Rutowski RL (1989) The function of the mating plug in the chalcedon checkerspot butterfly. Anim Behav 38:154–162

Downes JA (1973) Lepidoptera feeding at puddle-margins, dung, and carrion. J Lepid Soc 27:89–99

Dunlap-Pianka H, Boggs CL, Gilbert LE (1977) Ovarian dynamics in heliconiine butterflies: programmed senescence versus eternal youth. Science 197:487–490

Engqvist L, Sauer KP (2001) Strategic male mating effort and cryptic male choice in a scorpionfly. Proc R Soc Lond B 268:729–735

Ferkau C, Fischer K (2006) Costs of reproduction in male *Bicyclus anynana* and *Pieris napi* butterflies: effects of mating history and food limitation. Ethology 112:1117–1127

Fischer K, Fiedler K (2001) Effects of adult feeding and temperature regime on fecundity and longevity in the butterfly *Lycaena hippothoe*. J Lepid Soc 54:91–95

Fischer K, O'Brien DM, Boggs CL (2004) Allocation of larval and adult resources to reproduction in a fruit-feeding butterfly. Funct Ecol 18:656–663

Frey D, Roman R, Messett L (2002) Dew-drinking by male monarch butterflies, *Danaus plexippus* (L.). J Lepid Soc 56:90–97

Gage MJG (1994) Associations between body size, mating pattern, testis size and sperm length across butterflies. Proc R Soc Lond B 258:247–254

He Y, Miyata T (1997) Variations in sperm number in relation to larval crowding and spermatophore size in the armyworm, *Pseudaletia separata*. Ecol Entomol 22:41–46

Hiroyoshi S (1995) Regulation of sperm quantity transferring to females at mating in the adult male of *Polygonia c-aureum* (Lepidoptera: Nymphalidae). Appl Entomol Zool 30:111–119

Hiroyoshi S, Mitsuhashi J (1999) Sperm reflux and its role in multiple mating in males of a butterfly *Polygonia c-aureum Linnaeus* (Lepidoptera: Nymphalidae). J Insect Physiol 45:107–112

Kaitala A, Wiklund C (1994) Polyandrous female butterflies forage for matings. Behav Entomol Sociobiol 35:385–388

Kaitala A, Wiklund C (1995) Female mate choice and mating costs in the polyandrous butterfly *Pieris napi* (Lepidoptera: Pieridae). J Insect Behav 8:355–363

Krenn HW (2010) Feeding mechanisms of adult Lepidoptera: structure, function, and evolution of the mouthparts. Annu Rev Entomol 53:307–327

LaChance LE, Richard RD, Ruud RL (1977) Movement of eupyrene sperm bundles from the testis and storage in the ductus ejaculatoris duplex of the male pink bollworm: effects of age, strain, irradiation, and light. Ann Entomol Soc Am 71:647–651

Lederhouse RC (1981) The effect of female mating frequency on egg fertility in the black swallowtail, *Papilio polyxenes asterius* (Papilionidae). J Lepid Soc 35:266–277

Lederhouse RC, Ayres MP, Scriber JM (1990) Adult nutrition affects male virility in *Papilio glaucus* L. Funct Ecol 4:743–751

Lewis Z, Wedell N (2007) Effect of adult feeding on male mating behaviour in the butterfly, *Bicyclus anynana* (Lepidoptera: Nymphalidae). J Insect Behav 20:201–213

Marshall LD (1985) Protein and lipid composition of *Colias philodice* and *C. eurytheme* spermatophores and their changes over time. J Res Lepid 24:21–30

Matsumoto K, Suzuki N (1995) The nature of mating plugs and the probability of reinsemination in Japanese Papilionidae. In: Scriber JM, Tsubaki Y, Lederhouse RC (eds) Swallowtail butterflies: their ecology and evolutionary biology. Scientific Publishers, Gainesville, pp 145–154

May PG (1985) A simple method for measuring nectar extraction rates in butterflies. J Lepid Soc 39:53–55

Mazer CL, Appel AG (2001) Water loss and desiccation tolerances of longwing butterflies (Lepidoptera: Nymphalidae). Environ Entomol 30:631–636

McNamara KB, Elgar MA, Jones TM (2009) Large spermatophores reduce female receptivity and increase male paternity success in the almond moth, *Cadra cautella*. Anim Behav 77:931–936

Muraoka K, Watanabe M (1994) A preliminary study of nectar production of the field cress, *Rorippa indica*, in relation to the age of its flowers. Ecol Res 9:33–36

Niihara S, Watanabe M (2009) Effects of saline intake until the second mating on the ejaculates and sperm number of male swallowtail butterfly, *Papilio xuthus* Linnaeus. Jpn J Appl Entomol Zool 53:91–96 (in Japanese)

Norris MJ (1936) The feeding-habits of the adult Lepidoptera *Heteroneura*. Trans R Entomol Soc Lond 85:61–90

Oberhauser KS (1988) Male monarch butterfly spermatophore mass and mating strategies. Anim Behav 36:1384–1388

Oberhauser KS (1989) Effects of spermatophores on male and female monarch butterfly reproductive success. Behav Ecol Sociobiol 25:237–246

Oberhauser KS (1997) Fecundity, lifespan and egg mass in butterflies: effects of male-derived nutrients and female size. Funct Ecol 44:166–175

Parker GA (1970) Sperm competition and its evolutionary consequences in insects. Biol Rev 45:525–567

Pivnick KA, McNeil JN (1987) Puddling in butterflies: sodium affects reproductive success in *Thymelicus lineola*. Physiol Entomol 12:461–472

Proshold FI (1991) Number of sperm bundles in the duplex of tobacco bud worms (Lepidoptera: Noctuidae) as a function of age. J Econ Entomol 84:1485–1491

Rawlins JE (1980) Thermoregulation by the black swallowtail butterfly, *Papilio polyxenes* (Lepidoptera: Papilionidae). Ecology 61:345–357

Reinhold K, Kurtz J, Engqvist L (2002) Cryptic male choice: sperm allocation strategies when female quality varies. J Evol Biol 15:201–209

Rönn JL, Katvala M, Arnqvist G (2008) Interspecific variation in ejaculate allocation and associated effects on female fitness in seed beetles. J Evol Biol 21:461–470

Rutowski RL (1984) Sexual selection and the evolution of butterfly mating behavior. J Res Lepid 23:125–142

Rutowski RL, Gilchrist GW (1986) Copulation in *Colias eurytheme* (Lepidoptera: Pieridae): patterns and frequency. J Zool Lond A 209:115–124

Rutowski RL, Newton M, Schaefer J (1983) Interspecific variation in the size of the nutrient investment made by male butterflies during copulation. Evolution 37:708–713

Rutowski RL, Gilchrist GW, Terkanian B (1987) Female butterflies mated with recently mated males show reduced reproductive output. Behav Ecol Sociobiol 20:319–322

Rutowski RL, Alcock J, Carey M (1989) Hilltopping in the pipevine swallowtail butterfly (*Battus philenor*). Ethology 82:244–254

Sasaki N, Watanabe M (2015) Alternative ejaculate allocation tactics in relation to male mating history of the swallowtail butterfly, *Papilio xuthus* L. (Lepidoptera: Papilionidae). J Res Lepid 48:1–8

Sasaki N, Konagaya T, Watanabe M, Rutowski RL (2015) Indicators of recent mating success in the pipevine swallowtail butterfly (*Battus philenor*) and their relationship to male phenotype. J Insect Physiol 83:30–36

Simmons LW (2001) Sperm competition and its evolutionary consequences in the insects. Princeton University Press, Oxford

Simmons LW, Siva-Jothy MT (1998) Sperm competition in insects: mechanisms and the potential for selection. In: Birkhead TR, Møller AP (eds) Sperm competition and sexual selection. Academic, San Diego, pp 341–434

Sims SR (1979) Aspects of mating frequency and reproductive maturity in *Papilio zelicaon*. Am Midl Nat 102:36–50

Sugawara T (1979) Stretch reception in the bursa copulatrix of the butterfly, *Pieris rapae crucivora*, and its role in behaviour. J Comp Physiol 130:191–199

Suzuki N, Okuda T, Shinbo H (1996) Sperm precedence and sperm movement under different copulation intervals in the silkworm, *Bombyx mori*. J Insect Physiol 42:199–204

Svärd L, Wiklund C (1989) Mass and production rate of ejaculates in relation to monandry/polyandry in butterflies. Behav Ecol Sociobiol 24:395–402

Swallow JG, Wilkinson GS (2002) The long and short of sperm polymorphisms in insects. Biol Rev 77:153–182

Takeuchi T (2012) Cost of reproduction in males of a satyrine butterfly *Lethe diana*. Physiol Entomol 37:171–176

Tschudi-Rein K, Benz G (1990) Mechanisms of sperm transfer in female *Pieris brassicae* (Lepidoptera: Pieridae). Ann Entomol Soc Am 83:1158–1164

Vahed K (1998) The function of nuptial feeding in insects: a review of empirical studies. Biol Rev 73:43–78

Vande Velde L, Damiens D, Van Dyck H (2011) Spermatophore and sperm allocation in males of the monandrous butterfly *Pararge aegeria*: the female's perspective. Ethology 117:645–654

Vane-Wright D (2003) Butterflies. The Natural History Museum, London

Watanabe M (1988) Multiple matings increase the fecundity of the yellow swallowtail butterfly, *Papilio xuthus* L., in summer generations. J Insect Behav 1:17–29

Watanabe M (1992) Egg maturation in laboratory-reared females of the swallowtail butterfly, *Papilio xuthus* L. (Lepidoptera: Papilionidae), feeding on different concentration solutions of sugar. Zool Sci 9:133–141

Watanabe M, Ando S (1993) Influence of mating frequency of lifetime fecundity in wild females of the small white *Pieris rapae* (Lepidoptera, Pieridae). Jpn J Entomol 61:691–696

Watanabe M, Hachisuka A (2005) Dynamics of eupyrene and apyrene sperm storage in ovipositing females of the swallowtail butterfly *Papilio xuthus* (Lepidoptera: Papilionidae). Entomol Sci 8:65–71

Watanabe M, Hirota M (1999) Effects of sucrose intake on spermatophore mass produced by male swallowtail butterfly *Papilio xuthus* L. Zool Sci 16:55–61

Watanabe M, Kamikubo M (2005) Effects of saline intake on spermatophore and sperm ejaculation in the male swallowtail butterfly *Papilio xuthus* (Lepidoptera: Papilionidae). Entomol Sci 8:161–166

Watanabe M, Kobayashi T (2006) Total sperm ejaculation in monandrous (*Papilio machaon*) and polyandrous (*P. xuthus*) swallowtail butterflies (Lepidoptera: Papilionidae) restricted to larval stage-derived nutrients. J Res Lepid 39:1–7

Watanabe M, Sato K (1993) A spermatophore structured in the bursa copulatrix of the small white *Pieris rapae* (Lepidoptera, Pieridae) during copulation and its sugar content. J Res Lepid 32:26–36

Watanabe M, Koizumi H, Suzuki N, Kiritani K (1988) Studies on ecology and behavior of Japanese black swallowtail butterflies. 7. Nectar of a glory tree, *Clerodendron trichotomum*, as a food resource of adults in summer. Ecol Res 3:175–180

Watanabe M, Wiklund C, Bon'no M (1998a) The effect of repeated matings on sperm numbers in successive ejaculates of the cabbage white butterfly *Pieris rapae* (Lepidoptera: Pieridae). J Insect Behav 11:559–570

Watanabe M, Wiklund C, Bon'no M (1998b) Ejaculation timing of eupyrene and apyrene sperm in the cabbage white butterfly *Pieris rapae* (Lepidoptera, Pieridae) during copulation. Entomol Sci 1:15–19

Watanabe M, Bon'no M, Hachisuka A (2000) Eupyrene sperm migrates to spermatheca after apyrene sperm in the swallowtail butterfly, *Papilio xuthus* L. (Lepidoptera: Papilionidae). J Ethol 18:91–99

Watt WB, Hoch PC, Mills SG (1974) Nectar resource use by *Colias* butterflies. Oecologia (Berl) 14:353–374

Wedell N (2005) Female receptivity in butterflies and moths. J Exp Biol 208:3433–3440

Wedell N, Cook PA (1998) Determinants of paternity in a butterfly. Proc R Soc Lond B 265:625–630

Wedell N, Cook PA (1999) Strategic sperm allocation in the small white butterfly *Pieris rapae* (Lepidoptera: Pieridae). Funct Ecol 13:85–93

Chapter 6
Avoidance of Sperm Competition in Males

Keywords Harassment • Lone male • Mating plug • Mimicking • Resource allocation • Spermatheca, sperm depletion • Sperm migration • Sperm mixing • Takeover

Although courtship displays differ between species, virgin females of any species seem to accept copulation easily whenever certain aspects of male performance are adequate. On the other hand, mated females frequently show the open-winged posture and raised abdomen stance to prevent access to courting males (Fig. 6.1). Many studies have clarified the significance of mate choice of mated females during mate-locating behaviour, with detection of male qualities such as age and size. After acceptance by the mated females, the males succeed in copulating but eventually encounter the severe problem that their mates have already stored enough sperm from previous males for fertilisation. Transferred sperm have to compete with the sperm of previous males inside the female reproductive organs. Because the sex ratio is around unity with female polyandry, the probability of encounter with already mated females must be high. Therefore, males might have to develop not a morphological but a behavioural or physiological device to avoid or to win the sperm competition. Sperm migration inside the female reproductive organs after copulation suggests that apyrene sperm may intercept the movement of the eupyrene sperm of the previous mated males. On the other hand, sperm always have a certain risk for competition with future males that will be accepted by the females. To inhibit insemination by the sperm of future males, a spermatophore mass is the only male tactic to delay copulation of the mated females. Consequently, strategic ejaculation by males might be evolved for rival and future males.

Competition with Stored Rival Sperm

Spermatophores persist in the bursa copulatrix of female butterflies throughout their lives, although their size declines because of absorption by the females. In polyandrous species, most females encountered by males in the field throughout their lifespan have generally mated more than once, and their reproductive organs, the bursa copulatrix and the spermatheca, contain spermatophores and both types of

M. Watanabe, *Sperm Competition in Butterflies*, Ecological Research Monographs,
DOI 10.1007/978-4-431-55945-0_6

Fig. 6.1 A male black-veined white, *Aporia crataegi*, attacking the female, although she shows a typical mate refusal posture

sperm, respectively. In spite of the number and mass of the spermatophores in the bursa copulatrix, sperm stored in the spermatheca are used in fertilisation and might reside for long periods, even for the life of females. Because a single mating provides sufficient sperm for a female's lifetime egg production, male re-mating frequency is likely to be an index of the risk of sperm competition affecting male investment in spermatogenesis, as well as nutrients. If sperm mixing or displacement occurs in females, an increase in sperm number transferred by the current male is advantageous in competition with both prior and future rival male ejaculates. With such an increased risk of sperm competition, to ensure optimum fertilisation males have to invest more in their ejaculate. Thus, when the mated females re-mate, sperm competition must be the outcome, and a male's fertilisation success is likely to depend, in part, upon the number of his sperm that arrive and remain in the spermatheca. Sperm competition should select for the males to increase the number of sperm transferred to the female (Parker 1982). However, studies on several species suggest that males decrease the number of sperm transferred with successive matings (Simmons 2001); that is, the sperm production of males is limited. Transferring a large number of sperm seems to be costly for males (Pitnick and Markow 1994).

When males mate the mated females, the males respond by acting directly on competing sperm, either by removing it with penile brushes and scrapers, as in Odonata (Waage 1979), by disabling it with seminal secretions, as in fruit flies (Harshman and Prout 1994), by flooding it with large numbers of their own sperm, as in crickets (Gage and Barnard 1996) or by producing sperm with greater competitiveness, as in nematodes (LaMunyon and Ward 1998). These actions are taken directly against the competing sperm inside the female sperm storage organs. However, the morphology of female reproductive organs in butterflies shows that males are not able to directly manipulate the sperm of previous males in the spermatheca, because the spermatheca is completely separate from the bursa copulatrix where the spermatophore is formed during copulation.

Sperm is not mixed in the spermatheca because of the cul-de-sac morphology (Watanabe et al. 2000); thus, sperm from the more recent mating might be the first to leave from the spermatheca for fertilisation, resulting in last-male priority. Simmons and Siva-Jothy (1998) reviewed the data on the mean and variance in P_2 (i.e., the proportion of eggs fertilised by the second male in a double-mating trial) values for available studies of sperm utilisation in insects including butterflies, and showed that in general the last male to mate with a female tends to have a higher rate of paternity than the first (e.g., mean P_2 is 0.86, 1.00 and 0.72 for the mocker swallowtail, *Papilio dardanus*, the alfalfa butterfly, *Colias eurytheme*, and the checkerspot butterfly, *Euphydryas editha*, respectively). Therefore, male butterflies seem to use non-nucleated apyrene sperm, which are less energetically costly to produce than nucleated sperm, to exploit the design of a female's reproductive organ system, which monitors the number of sperm in storage and then prevents fertilisation by eupyrene sperm from former mates, thus reducing the potential for sperm competition.

Competition with Future Ejaculates

Postcopulatory sexual selection is important for males, particularly in polyandrous species, because males do not always fertilise all the eggs laid by his mates. If the first male's sperm do not fertilise all offspring produced after a subsequent mating, it might be in his interest to prevent female re-mating. There is a sperm precedence, in that the last male to mate fathered subsequent offspring, probably not in all but in most butterfly species. Then, the reproductive success of the male is not increased with the number of matings throughout his lifespan. Sperm transferred by the male have to compete for fertilisation of the eggs in the reproductive organs of females mated multiply even after successful copulation. On the other hand, mated females of polyandrous butterflies frequently display a period of nonreceptivity following the mating because of the ejaculate transferred by the male. Mating plugs, the sphraga, formed by males during copulation can be remarkably large and elaborate to prevent further mating of females in some butterfly species, most of which are considered monandrous species. However, such a mating plug has a transient effect (Dickinson and Rutowski 1989). In the chalcedon checkerspot butterfly, *Euphydryas chalcedona*, females from which the plug was removed experimentally are just as likely to reject courting males as females with an intact mating plug, indicating that other factors must affect female receptivity. The number of apyrene sperm in the spermatheca also increases the refractory period of female re-mating (Cook and Wedell 1999).

Transferring a larger spermatophore would be a greater fertilisation success, probably because more sperm are transferred, although a large spermatophore does not always include many sperm in some species, as reported by Cook and Wedell (1996). The benefits of producing a large spermatophore have been discussed in many lepidopteran species from the male perspective (Gwynne 2008). Then, the spermatophore represents a male's mating effort, because a large spermatophore can increase the female refractory period (Sugawara 1979), resulting in more eggs being fertilised by the male's sperm. In addition, the spermatophore functions as a paternal investment, because nutrients contained in the spermatophore are used by a female to increase her longevity (Boggs and Watt 1981) and reproductive output (Watanabe 1988). Because the male's ability to produce a spermatophore is heritable (Wedell 2006), females gain an indirect benefit by choosing sperm from the male that transferred a larger spermatophore.

The spermatheca seems to be fully filled by the sperm of the first male, when the copulation is normally terminated. The spermatheca of the lepidopteran females may have a restricted sperm storage capacity (Watanabe and Hachisuka 2005). LaMunyon (2000) reported that the storage capacity of the spermatheca of the tobacco budworm, *Heliothis virescens*, is approximately the mass of one ejaculate. Then, an alternative to the first male sperm priority hypothesis is that females that received a smaller spermatophore from the second male did not allow the second male's sperm to enter the spermatheca, and thus did not allow sperm mixing. Accordingly, the P_2 value must exhibit a bimodal distribution, probably because sperm migration from spermatophore to spermatheca is dependent largely on the female's musculature (Tschudi-Rein and Benz 1990). Curril and LaMunyon (2006) suggested that females of the bella moth, *Utetheisa ornatrix*, shunt the sperm of unwanted males to different organs rather than to the spermatheca.

Sasaki et al. (2015) showed that P_2 values of the swallowtail butterfly, *Papilio xuthus*, exhibited a bimodal distribution with peaks at 0% and 100%, mainly depending on the spermatophore size. Watanabe and Sasaki (2010) clarified that sperm ejection by female swallowtail butterflies occurred when the female received a larger spermatophore from the second male, and suggested that sperm ejection is the mechanism by which paternity is biased towards preferred males. Such female behaviour has been termed cryptic female choice. As pointed out by Watanabe et al. (2000), once sperm ejection occurs, almost all sperm of the first male in the spermatheca loses, and the remaining sperm are placed in the background of the spermatheca by the second male's sperm. Consequently, most eggs laid after the second mating will have been fertilised by the second male's sperm.

Most eggs can be fertilised by only one male chosen by the female, because male ejaculates usually contain far more sperm than needed for the fertilisation of all eggs. However, if sperm are lost by use for fertilisation, a longer interval duration between matings could result in higher second male priority. Although a single spermatozoon may be cheap for production, the ejaculation is costly to the male, because the male has to accumulate and transfer large numbers of sperm (Bissoondath and Wiklund 1997). It is found that 2 to 3 days of exclusive feeding on nectar is needed for the recovery of the spermatophore mass. Therefore, males should not re-mate within 1 to 2 days after the previous mating, when the probability of future matings is high.

Prolonged Copulation

Although mating costs time, mate guarding is considered to have evolved in the context of sexual selection, to protect a male's sperm investment from conspecific males competing for fertilisation. Oberhauser (1989) showed that monarch butterflies spend several hours in copula (mainly at night). However, many butterfly species spend much less time in copula during the daytime. Once mating has started, the male decides when to terminate copulation (Wickman 1985). Mated males of the cabbage white, *Pieris rapae*, increase copula duration but reduce spermatophore mass with the number of matings. The former increase suggests that such prolonged matings are probably the result of the mating history of the males that may accumulate spermatophore material in their accessory glands (Watanabe et al. 1998). On the other hand, males start to transfer immediately after the onset of copulation, and continuously transfer accessory gland materials (Watanabe and Sato 1993). Therefore, the prolonged copulation that continues after the termination of transferring the materials must have a function other than spermatophore transfer.

Prolonged copulation involves either precopulatory or postcopulatory mate-guarding behaviour by males. For butterflies, prolonged copulations are observed and have been explained as a male in-copula guarding strategy: his body acts as a nuptial plug and impedes additional copulations of the female. Thus, prolonged matings may have evolved at least partially as a result of the sexual selective pressure to reduce sperm competition before oviposition. For the pale clouded yellow, *Colias erate*, in Japan (Fig. 6.2), females are receptive and may mate multiply during their lifespan, making it easier for males to find mates, although the operational sex ratio is male biased. Throughout the day, males search for females which feed on the nectar of the clovers, *Trifolium repens* and *T. pratense*, the dominant species of grassland vegetation (Watanabe and Nakanishi 1996). When males fly around these habitats, they can easily encounter either copulating pairs or lone females. Lone males usually try to harass the copulating pairs as well as to mate with unreceptive females. One can commonly see many males pursuing a female above the grassland in summer season. When copulating pairs are harassed by single males, the copula duration is increased for more than 2 h (Watanabe et al. 1997). On the other hand, when the lone male encounters a lone female, he usually tries persistently to court her.

Mating males actually prolong copulation to guard their mates against the harassment by lone males. Because the frequency of prolonged copulations is affected by operational sex ratio as well as female status, prolonged copulation could be a male strategy to prevent sperm competition by subsequent matings, particularly at high male densities (Cordero 1990). In other words, time spent in copula may be related to the number of available females.

The insemination of a structure such as a mating plug must mechanically elicit the female to be nonreceptive soon after the copulation. The role of the prolonged copulation is similar to the postcopulatory strategy, because the male himself is

Fig. 6.2 A male pale clouded yellow, *Colias erate*, feeding on the eastern daisy fleabane

analogous to a mating plug. However, the mating plug can leave the male free to search for, and mate with, other females, whereas in prolonged copulation there is always a risk that the female will mate again before oviposition because the end of prolonged copulation means the end of the direct guarding. This risk could be increased by a small spermatophore, which elicits the repeat mating of the female in most butterflies. Old males of the pale clouded yellow, *Colias erate*, ejaculate a smaller spermatophore than younger ones. Then, harassing males strongly discriminate against old males copulating, probably because of scale loss, and the old males are apt to show prolonged copulation, irrespective of the degree of harassment (Watanabe et al. 1997). A female takeover by the single males harassing occurs mainly against old males when the females would accept pairing with the interloper.

On the other hand, it is possible that females actually avoid males at times. Interference by males when females want to lay eggs has been seen to drive females from the flight areas (Shapiro 1970). To avoid harassment by males and excessive matings, most females might have adopted mate refusal postures to reduce their harassment. In the pale clouded yellow, *Colias erate*, andromorphs rather than albas could avoid excessive time-consuming matings and male harassment by mimicking males (Watanabe et al. 1997). Then, males tend to approach lone alba females.

However, because the majority of albas had mated previously, to elicit a re-mating, a long period of courtship behaviour by the males is required.

Strategic Ejaculation

Ejaculate production including sperm is frequently associated with costs, and males must tailor their ejaculates to maximise their reproductive success, probably allocating their ejaculates strategically. Because females mate multiply throughout their lifespan, both spermatophore size and sperm number have an impact on the reproductive success of butterfly males. However, intake of most nutrients in the spermatophore produced by the male is available only during larval feeding, and it is not easy for nectar-feeding butterflies to replenish these materials after eclosion. Indeed, longevity of the males of the green-veined white, *Pieris napi*, decreased with increased number of matings, probably because of undernourishment (Ferkau and Fischer 2006). A significant increase in spermatophore mass of virgin males over time after emergence has been reported in the swallowtail butterfly, *Papilio xuthus* (Watanabe and Hirota 1999) and in the cabbage white, *Pieris rapae* (Wedell and Cook 1999a). After the first copulation, the spermatophore mass of the mated males was less: the spermatophore from the second copulation 1 day after the first copulation was approximately half the size of the first in the swallowtail butterfly without feeding on nectar. Mated males then increased their spermatophore mass with each additional day of rest, indicating that ejaculate production in males is associated with costs. Therefore, resource allocation among matings must be critical for lepidopteran males.

Risk and intensity of sperm competition have been regarded as important factors to determine the amount of male investment during mating (Parker et al. 1996), and plasticity of the amount of male investment at the matings has been clarified for several species. For example, the number of sperm transferred by male crickets increased with the operational sex ratio (Gage and Barnard 1996). Males of the Indian meal moth, *Plodia interpunctella*, transferred many sperm when they mated with large or previously mated females (Cook and Gage 1995). On the second mating of the male cabbage white, *Pieris rapae*, both the mass of the spermatophore and the number of sperm increase with time since the first mating (Wedell and Cook 1999a). However, differences in ejaculate allocation patterns across species are poorly understood, although the degree of polyandry that should strongly affect the degree of sperm competition differs among species (Cook and Wedell 1996).

Theoretical studies had indicated that the optimal male investment at a mating depends on the quantity of resources available to the male for reproduction (Engqvist and Reinhold 2007). In Lepidoptera, sperm are released from the testis and transported via the vas deferens to the duplex where they are stored until mating. During copulation, most sperm stored in the duplex are transferred to the female; however, a part of these are moved back to the vas deferens and then restored later in the duplex (Hiroyoshi and Mitsuhashi 1999). This finding supports that males have an ability to regulate the number of sperm transferred during copulation, as shown by the example previously mentioned. These facts enable us to investigate not only how male butterflies produce sperm before matings, but also how the amount of sperm transferred changes with the amount of sperm produced, by counting the number of sperm in the spermatophore immediately after mating and in the male sperm storage organs (the duplex and the vas deferens). In the cabbage white, *Pieris rapae*, the rate of increase in eupyrene sperm numbers is higher after the first mating, and the difference in rate of increase between eupyrene and apyrene sperm may be the result of different probabilities of virgin and non-virgin males obtaining future matings (Wedell and Cook 1999a). Males seem to be able to decide how many sperm to ejaculate at a given mating. In the swallowtail butterfly, *Papilio xuthus*, Sasaki and Watanabe (2015) showed a negative correlation between the ratio of sperm transferred to the female and the number of sperm produced in the virgin males. The virgin males seem to have an ability to keep some sperm for further matings, whereas the mated males transfer most of the sperm in their sperm storage organs at the their second mating, irrespective of the number of sperm stored (Fig. 6.3). Thus, the ability of males to store sperm in the duplex provides a mechanism for strategic allocation of sperm between matings.

Wedell and Cook (1999a) stated two possible explanations for the increase with time in the number of eupyrene sperm ejaculated in mated males. The first explanation is that sperm release from the testis is stimulated to increase after a male's first mating, although the acceleration of eupyrene sperm release has not been reported. The second explanation is that sperm is released from the testis at the same rate throughout the lifetime, but a male retains more eupyrene sperm in his duplex after his first mating that might be ejaculated on his second mating. Therefore, by retaining eupyrene sperm in the duplex, males ensure that high numbers are available for their second mating, which is more likely to be with mated females. In other word, a decrease of ejaculate size over successive matings must be adaptive in species where female re-mating frequency is low. Reinhold et al. (2002) suggested that if young females are predominant early in each discrete generation, males should transfer the larger ejaculate in early copulations, because fecundity of females and the probability of achieving additional matings decreases over time, and sperm competition intensity increases over time.

In polyandrous species with relatively long adult longevity, when sperm or any other investment in the mating might limit male reproductive success, males could be expected to choose cryptically by varying their investment. For summer generations of the Japanese swallowtail butterflies, adults are continuously seen because of the long suitable flying season and their relatively long longevity (Watanabe and Kobayashi 2006). Consequently, there is a complex age structure of females with much variation in fecundity and mating status throughout the flying season. Therefore, males of polyandrous swallowtail species such as *Papilio xuthus* should allocate a more equal amount of ejaculate resources over successive matings. However, because the simplex of the male butterflies just after mating is almost empty, these males have to improve their ability to produce spermatophore substances after the first mating so that they could recover their spermatophore mass

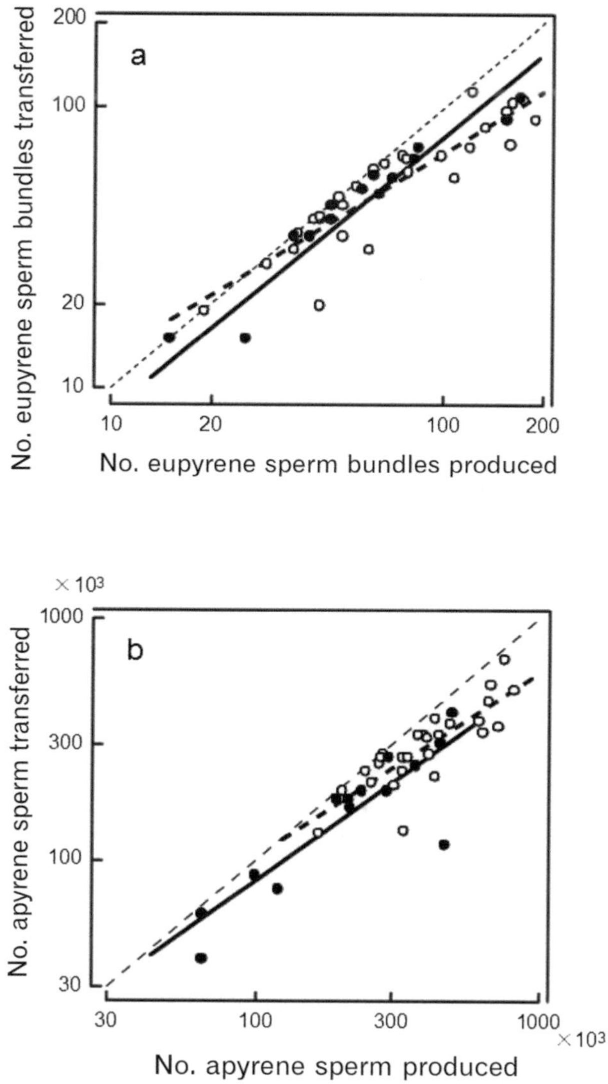

Fig. 6.3 (**a**) Relationship between the number of eupyrene sperm bundles produced and the number of eupyrene sperm bundles transferred for virgin males of the swallowtail butterfly, *Papilio xuthus* (*open circles* with *dotted line*; $\log Y = 0.39 + 0.73 \log X$, $r^2 = 0.82$, $n = 29$) and for previously mated males (*solid circles* with *solid line*; $\log Y = -0.07 + 0.98 \log X$, $r^2 = 0.88$, $n = 15$). (**b**) Relationship between the number of apyrene spermatozoa produced and the number of eupyrene sperm bundles transferred for virgin males of the swallowtail butterfly, *Papilio xuthus* (*open circles* with *dotted line*; $\log Y = 1.23 + 0.76 \log X$, $r^2 = 0.71$, $n = 29$) and for mated males (*solid circles* with *solid line*; $\log Y = 0.62 + 0.86 \log X$, $r^2 = 0.76$, $n = 15$) (Sasaki and Watanabe 2015)

quickly before the next mating. The longer the time before a male re-mates, the more sperm with the ejaculates might have accumulated in the duplex. Then, when the male becomes older, more sperm could be delivered on his second mating (Sasaki and Watanabe 2015).

Although males have been shown to ejaculate more sperm when mating with females providing higher fecundity returns, males are often sensitive to cues indicating the level of sperm competition and adjust their ejaculate expenditure accordingly (Wedell and Cook 1999b). Female quality as well as the stochastic nature of male mating success might influence the optimal sperm allocation

Table 6.1 Description of conditions assumed in the four simulation series used to examine the evolutionarily stable sperm allocation strategies

Simulation series	Variation in female quality	Males can detect female quality	Sperm refilling
I	–	–	–
II	+	–	–
III	+	+	–
IV	+	+	+

Source: Reinhold et al. (2002)

strategy. Thus, males could reduce their ejaculate expenditure, particularly when facing the risk of sperm competition with the sperm of related males (Lewis and Wedell 2009).

Assuming that fertilisation efficiency is equal among the swallowtail butterflies, the number of eupyrene spermatozoa required to inseminate females should be roughly the same (Watanabe and Kobayashi 2006), suggesting that males of *Papilio xuthus* (polyandrous) might produce more eupyrene sperm than *Papilio machaon* (monandrous). This difference is possible because, under sperm competition in the polyandrous species, the production of high numbers of sperm is advantageous when sperm from different males mix in the female's spermatheca. In addition, the males of the polyandrous swallowtail butterfly seem to be capable of storing sperm in their duplex, allowing them to increase the number of sperm on their second mating, as well as to reserve sperm for yet another mating.

Recent studies suggest that male butterflies have the ability to regulate the number of sperm transferred during copulation. Assuming an occurrence of sperm mixing in females, a limited ejaculate volume in males and a stochastic male mating frequency, Reinhold et al. (2002) examined how males should allocate their sperm over successive copulations in four simulation series that differed in the following characteristics (Table 6.1):

1. In the assumed variation in female quality
2. In male ability to recognise their variation
3. In male refilling of sperm stores

The simulations predicted that males should invest more sperm in the first copulation than subsequent copulations and more sperm in high-quality females than in low-quality females. However, with increasing variance in female quality, males could conserve more sperm for later copulations.

Production of sperm rather than a spermatophore requires much nitrogen (Boggs and Gilbert 1979), and it is difficult to replenish nitrogen after eclosion in nectar-feeding butterflies. A high sperm number must achieve higher fertilisation success, resulting in increased paternity when sperm from different males mix in the female's spermatheca, although the effect of numerous sperm decreases with degree of immiscibility of sperm from different males (Parker et al. 1997). In Lepidoptera, a bimodal distribution of P_2 values with peaks near 0 and 1 is common

(Simmons 2001), indicating that sperm mixing rarely occurs, Therefore, the advantage of high sperm number might be relatively small in butterflies. In fact, males of the swallowtail butterfly, *Papilio xuthus*, may be exposed to strong selection pressures to maintain spermatophore size, and then they may have evolved to produce fewer eupyrene spermatozoa but larger spermatophores rather than the other *Papilio* species, such as *P. machaon*, a monandrous species. Males of polyandrous species might have a mechanism to transfer sperm more prudently than monandrous males (Reinhold et al. 2002). Hiroyoshi (1995) stated that the sperm-saving system of the males of the comma butterfly, *Polygonia c-aureum*, may serve to enhance performance during multiple mating, in view of the risk of sperm depletion caused by multiple matings of the male itself. Cook and Gage (1995) stated that the male Indian meal moth, *Plodia interpunctella*, may reduce eupyrene sperm numbers to older virgin females because aged females are of lower reproductive value and do not contain sperm.

References

Bissoondath CJ, Wiklund C (1997) Effect of male body size on sperm precedence in the polyandrous butterfly *Pieris napi* L. (Lepidoptera: Pieridae). Behav Ecol 8:518–523

Boggs CL, Gilbert LE (1979) Male contribution to egg production in butterflies: evidence for transfer of nutrients at mating. Science 206:83–84

Boggs CL, Watt WB (1981) Population structure of pierid butterflies. IV. Genetic and physiological investment in offspring by male *Colias*. Oecologia (Berl) 50:320–324

Cook PA, Gage MJG (1995) Effects of risks of sperm competition on the numbers of eupyrene and apyrene sperm ejaculated by moth *Plodia interpunctella* (Lepidoptera: Pyralidae). Behav Ecol Sociobiol 36:261–268

Cook PA, Wedell N (1996) Ejaculate dynamics in butterflies: a strategy for maximizing fertilization success? Proc R Soc Lond B 263:1047–1051

Cook PA, Wedell N (1999) Non-fertile sperm delay female remating. Nature (Lond) 397:486

Cordero A (1990) The adaptive significance of the prolonged copulations of the damselfly, *Ischnura graellsii* (Odonata: Coenagrionidae). Anim Behav 40:43–48

Curril IM, LaMunyon CW (2006) Sperm storage and arrangement within females of the arctiid moth *Utetheisa ornatrix*. J Insect Physiol 52:1182–1188

Dickinson JL, Rutowski RL (1989) The function of the mating plug in the chalcedon checkerspot butterfly. Anim Behav 38:154–162

Engqvist L, Reinhold K (2007) Sperm competition games: optimal sperm allocation in response to the size of competing ejaculates. Proc R Soc Lond B 274:209–217

Ferkau C, Fischer K (2006) Costs of reproduction in male *Bicyclus anynana* and *Pieris napi* butterflies: effects of mating history and food limitation. Ethology 112:1117–1127

Gage AR, Barnard CJ (1996) Male crickets increase sperm number in relation to competition and female size. Behav Ecol Sociobiol 38:349–353

Gwynne DT (2008) Sexual conflict over nuptial gifts in insects. Annu Rev Entomol 53:83–101

Harshman LG, Prout T (1994) Sperm displacement without sperm transfer in *Drosophila melanogaste*r. Evolution 48:758–766

Hiroyoshi S (1995) Regulation of sperm quantity transferring to females at mating in the adult male of *Polygonia c-aureum* (Lepidoptera: Nymphalidae). Appl Entomol Zool 30:111–119

Hiroyoshi S, Mitsuhashi J (1999) Sperm reflux and its role in multiple mating in males of a butterfly *Polygonia c-aureum Linnaeus* (Lepidoptera: Nymphalidae). J Insect Physiol 45:107–112

LaMunyon CW (2000) Sperm storage by females of the polyandrous noctuid moth *Heliothis virescens*. Anim Behav 59:395–402

LaMunyon CW, Ward S (1998) Larger sperm outcompete smaller sperm in the nematode *Caenorhabditis elegans*. Proc R Soc Lond Ser B 266:263–267

Lewis Z, Wedell N (2009) Male moths reduce sperm investment in relatives. Anim Behav 77:1547–1550

Oberhauser KS (1989) Effects of spermatophores on male and female monarch butterfly reproductive success. Behav Ecol Sociobiol 25:237–246

Parker GA (1982) Why are there so many tiny sperm? Sperm competition and the maintenance of two sexes. J Theor Biol 96:281–294

Parker GA, Ball MA, Stockley P, Gage MJG (1996) Sperm competition games: individual assessment of sperm competition intensity by group spawners. Proc R Soc Lond B 263:1291–1297

Parker GA, Ball MA, Stockley P, Gage MJG (1997) Sperm competition games: a prospective analysis of risk assessment. Proc R Soc Lond B 264:1793–1802

Pitnick S, Markow TA (1994) Male gametic strategies: sperm size, testes size, and the allocation of ejaculate among successive mates by the sperm-limited fly *Drosophila pachea* and its relatives. Am Nat 143:785–819

Reinhold K, Kurtz J, Engqvist L (2002) Cryptic male choice: sperm allocation strategies when female quality varies. J Evol Biol 15:201–209

Sasaki N, Watanabe M (2015) Alternative ejaculate allocation tactics in relation to male mating history of the swallowtail butterfly, *Papilio xuthus* L. (Lepidoptera: Papilionidae). J Res Lepid 48:1–8

Sasaki N, Todoriki S, Watanabe M (2015) Effect of sperm ejection by females on male fertilization success in the swallowtail butterfly, *Papilio xuthus* L. (Lepidoptera: Papilionidae). J Res Lepid 48:39–44

Shapiro AM (1970) The role of sexual behavior in density-related dispersal of pierid butterflies. Am Nat 104:367–372

Simmons LW (2001) Sperm competition and its evolutionary consequences in the insects. Princeton University Press, Oxford

Simmons LW, Siva-Jothy MT (1998) Sperm competition in insects: mechanisms and the potential for selection. In: Birkhead TR, Møller AP (eds) Sperm competition and sexual selection. Academic, San Diego, pp 341–434

Sugawara T (1979) Stretch reception in the bursa copulatrix of the butterfly, *Pieris rapae crucivora*, and its role in behaviour. J Comp Physiol 130:191–199

Tschudi-Rein K, Benz G (1990) Mechanisms of sperm transfer in female *Pieris brassicae* (Lepidoptera: Pieridae). Ann Entomol Soc Am 83:1158–1164

Waage JK (1979) Dual function of the damselfly penis: sperm removal and transfer. Science 203:916–918

Watanabe M (1988) Multiple matings increase the fecundity of the yellow swallowtail butterfly, *Papilio xuthus* L., in summer generations. J Insect Behav 1:17–29

Watanabe M, Hachisuka A (2005) Dynamics of eupyrene and apyrene sperm storage in ovipositing females of the swallowtail butterfly *Papilio xuthus* (Lepidoptera: Papilionidae). Entomol Sci 8:65–71

Watanabe M, Hirota M (1999) Effects of sucrose intake on spermatophore mass produced by male swallowtail butterfly *Papilio xuthus* L. Zool Sci 16:55–61

Watanabe M, Kobayashi T (2006) Total sperm ejaculation in monandrous (*Papilio machaon*) and polyandrous (*P. xuthus*) swallowtail butterflies (Lepidoptera: Papilionidae) restricted to larval stage-derived nutrients. J Res Lepid 39:1–7

Watanabe M, Nakanishi Y (1996) Population structure and dispersals of the sulfur butterfly *Colias erate* (Lepidoptera: Pieridae) in an isolated plain located in a cool temperate zone of Japan. Jpn J Entomol 64:17–29

Watanabe M, Sasaki N (2010) Pattern of sperm storage and migration in the reproductive tract of the swallowtail butterfly *Papilio xuthus*: cryptic female choice after second mating. Physiol Entomol 35:328–333

Watanabe M, Sato K (1993) A spermatophore structured in the bursa copulatrix of the small white *Pieris rapae* (Lepidoptera, Pieridae) during copulation and its sugar content. J Res Lepid 32:26–36

Watanabe M, Nakanishi Y, Bon'no M (1997) Prolonged copulation and spermatophore size ejaculated in the sulfur butterfly, *Colias erate* (Lepidoptera: Pieridae) under selective harassments of mated pairs by conspecific lone males. J Ethol 15:45–54

Watanabe M, Wiklund C, Bon'no M (1998) The effect of repeated matings on sperm numbers in successive ejaculates of the cabbage white butterfly *Pieris rapae* (Lepidoptera: Pieridae). J Insect Behav 11:559–570

Watanabe M, Bon'no M, Hachisuka A (2000) Eupyrene sperm migrates to spermatheca after apyrene sperm in the swallowtail butterfly, *Papilio xuthus* L. (Lepidoptera: Papilionidae). J Ethol 18:91–99

Wedell N (2006) Male genotype affects female fitness in a paternally investing species. Evolution 60:1638–1645

Wedell N, Cook PA (1999a) Strategic sperm allocation in the small white butterfly *Pieris rapae* (Lepidoptera: Pieridae). Funct Ecol 13:85–93

Wedell N, Cook PA (1999b) Butterflies tailor their ejaculate in response to sperm competition risk and intensity. Proc R Soc Lond B 266:1033–1039

Wickman P-O (1985) Male determined mating duration in butterflies? J Lepid Soc 39:341–342

Chapter 7
Sperm Manipulation in Females

Keywords Bimodal distribution • Cryptic female choice • Fat body • Fecundity • Hibernation • Mate avoidance • Mating order • P_2 value • Reproductive diapause • Sperm depletion

For success, a female butterfly has to develop eggs, to mate, to be inseminated and then to place those eggs in sites where there is the greatest likelihood of larval survival. Because early-instar larvae have little choice as to what they eat and depend on the judgement of the female to place eggs, females are highly selective in their choice of larval host plants. In searching for an egg-laying site, females have to respond to several specific demands that will affect the survival of their offspring. Although sufficient host plants are available for oviposition, the fitness of the females that mated with more than one male could be increased, because the eggs have been fertilised with the sperm of various males, probably having adaptive traits to survive in different types of habitats. Therefore, females are apt to mate more than once and to exploit an excess of sperm transferred, such as absorption of apyrene sperm. In addition to the adaptive oviposition process for suitable host plants, sexual selection must occur. Females could prefer the sperm of males that increase fitness of the females, for example, males that produce large spermatophores. When the signals for good males are not recognised by females until copulation, females have to assess their sperm quality as well as the spermatophore mass and decide to use the fertilisation products after copulation. Consequently, females have developed a technique to select suitable sperm in their reproductive organs, the so-called cryptic female choice.

Females of Monandrous Species

There is a positive relationship between the degree of polyandry and ejaculate size (Svärd and Wiklund 1989). In polyandrous species, females that receive a small initial spermatophore re-mate sooner (Wedell and Cook 1999) and may have reduced fecundity (Oberhauser 1997). On the other hand, in monandrous species, receiving a sufficiently large spermatophore at the first and the only single mating is also likely to be a significant life history issue. Examination of the effects of male

M. Watanabe, *Sperm Competition in Butterflies*, Ecological Research Monographs,
DOI 10.1007/978-4-431-55945-0_7

investment strategies on female mating frequency in the monandrous species, comparing with those in the polyandrous species, can provide insight into the processes that have limited the current female mating frequency. Watanabe and Kobayashi (2006) reported that virgin males transferred a spermatophore and accessory substances, corresponding to about 2.4 % and 2.1 % of male weight at eclosion, in the polyandrous (*Papilio xuthus*) and monandrous (*Papilio machaon*) swallowtail butterfly, respectively. Their spermatophores contained 41 and 120 eupyrene sperm bundles, and 247,000 and 202,000 apyrene spermatozoa, for *P. xuthus* and *P. machaon*, respectively. For their females, the mean lifetime mating frequency is 3 and 1 for *P. xuthus* and *P. machaon*, respectively (Watanabe and Nozato 1986). Assuming that every female attempted to store a sufficient number of eupyrene spermatozoa for insemination of her eggs, the total number of eupyrene spermatozoa in relationship to her fecundity must be similar among species, because the number of eggs in the ovaries was roughly the same. As a female of *P. xuthus* mates with three virgin males during her lifespan, she must receive 123 ($= 41 \times 3$) eupyrene sperm bundles. This number is comparable with the number received at a single mating by a female of *P. machaon*, although the eupyrene spermatozoon size of *P. machaon* is smaller than that of *P. xuthus*. On the other hand, however, a female of *P. xuthus* mating three times receives 750,000 apyrene spermatozoa.

Because the females of monandrous species mate with one male, the number of spermatophores received seems not to affect her reproductive output, indicating that the few nutrients from the spermatophore must be used for egg development in the females of monandrous species (McNamara et al. 2009). Females of the common yellow swallowtail *Papilio machaon* do not use the spermatophore and accessory substances for either energy or egg production, and they remain monandrous throughout their lifespan. Watanabe and Kobayashi (2006) pointed out that *P. machaon* had larger fat bodies in their abdomens after eclosion than did *P. xuthus* females. This finding corresponds with the body weights of the males. Males of *P. machaon* also have clearly more fat body than male *P. xuthus*, even though their wing sizes are similar. Therefore, non-virgin females of *P. machaon* must have required no further ejaculate from males than the mated females of *P. xuthus*, and might avoid their conspecifics or refuse further mating when the conspecific males are encountered. Because only virgin females permit copulation, the operational sex ratio might be biased towards males throughout the flying season. Consequently, the probability of mating in the males may be low in the field. In addition, the probability of re-mating by the mated males must be towards zero if females are able to detect mated males. Then, males of *P. machaon*, particularly virgins, might increase the number of eupyrene spermatozoa as well as the maximum ejaculation mass as much as possible when they copulate. Accordingly, sperm size has to be decreased.

Although the number of apyrene spermatozoa in the polyandrous swallowtail butterfly *Papilio xuthus* significantly decreases for any second mating, the lifetime production of apyrene spermatozoa is higher than that in the monandrous swallowtail butterfly, *P. machaon*, particularly in relationship to the number of eupyrene spermatozoa (Watanabe and Kobayashi 2006). The fact that apyrene spermatozoa migrate earlier than eupyrene spermatozoa from spermatophore to spermatheca following copulation suggests that apyrene spermatozoa for the second mating can fill the spermatheca with eupyrene spermatozoa of the first mating to prevent fertilisation (Watanabe et al. 2000). Therefore, it may be advantageous for the swallowtail butterfly, *Papilio xuthus*, males to provide polyandrous females with a certain number of large apyrene spermatozoa at each mating to effect sperm competition, rather than the monandrous swallowtail butterfly, *P. machaon*.

Intact spermatophores have been observed in the bursa copulatrix of aged females of monandrous species in the field such as the small copper, *Lycaena phlaeas* (Watanabe and Nishimura 2001), suggesting that these females do not use male ejaculate to obtain the nutrients necessary for reproductive success (Fig. 7.1). As shown in Fig. 7.2, in age FF females, the average volume of the spermatophore is about 0.05 mm^3, which occupied about 31 % of the volume of the bursa copulatrix. There is no significant difference in the size of the spermatophores among females of age FF to BB. Among females of these four age classes, the spermatophore occupies about 40 % of the bursa copulatrix. In age BBB females, the spermatophore size significantly decreases, comparing with the spermatophore size of age F, age B and age BB females, but still occupies about 45 % of the bursa copulatrix. Bissoondath and Wiklund (1995) stated that both relative ejaculate mass and protein content in the spermatophore are low in monandrous species.

Although spermatophore mass is not necessarily correlated with the amount of sperm (Oberhauser 1992), females of the speckled wood butterfly, *Pararge aegeria*, that typically mate only once, received a very different size of spermatophore (Vande Velde et al. 2011). The females might not discriminate to accept copulation

Fig. 7.1 A male of the small copper, *Lycaena phlaeas*, feeding on the eastern daisy fleabane

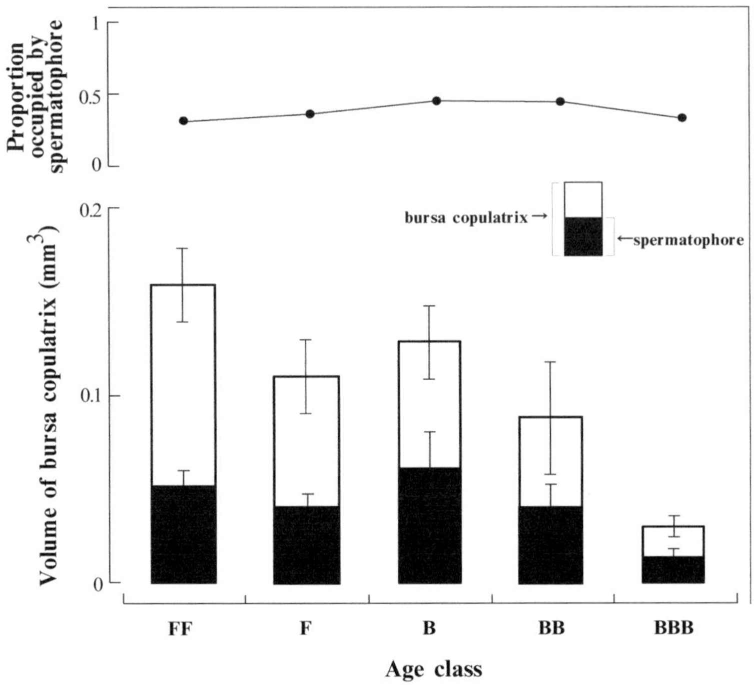

Fig. 7.2 Change in the volume of spermatophore in the bursa copulatrix and its proportion in wild monandrous females of the small copper, *Lycaena phlaeas*, for five age classes (*FF*, *F*, *B*, *BB*, *BBB*). Each *bar* represents SE. (From Watanabe and Nishimura 2001)

between virgin and non-virgin males that would provide a small spermatophore, sometimes with a low number of eupyrene sperm. Then, re-mating after having received a small spermatophore has been recognised as an extra mating in such monandrous species, as well as in polyandrous species. Such a small spermatophore might be insufficient to stimulate the stretch receptors in the bursa copulatrix of the female to induce a long refractory period, even in the monandrous species.

Watanabe and Nishimura (2001) found that most females (94 %) of the small copper, *Lycaena phlaeas*, in the field had a single spermatophore in the bursa copulatrix irrespective of their age class (Table 7.1). A few polygamous females (6 %) had two spermatophores, one of normal size lying near to the site of sperm migration in the bursa copulatrix and the other being very small or consisting of fragments, which might be transferred by old males or by recently mated males. That is, when the first spermatophore is too small for females to avoid recopulation, females must be exceptionally polygamous. Elzinga et al. (2011) also reported that monandrous females of Naryciinae species were able to re-mate after a nonsuccessful copulation. Because spermatophore size has little effect on female reproductive output in the monandrous species, the mechanism of paternal investment by males might differ between polyandrous and monandrous species.

Table 7.1 Frequency of spermatophore distribution in the bursa copulatrix of females of the small copper, *Lycaena phlaeas*, caught in the wild for five age classes (FF–BBB) in the summer generations of 1995 and 1996

Year	Number of spermatophores present	Age class				
		FF	F	B	BB	BBB
1995	0	5	2	0	0	0
	1	29	16	7	1	0
	2	2	2	0	0	0
1996	0	2	1	0	0	0
	1	3	9	6	8	4
	2	0	0	1	0	1

Source: Watanabe and Nishimura (2001)

Odendaal (1989) reported that the lifetime mating number of the pipevine swallowtail, *Battus philenor*, in Texas was 1.4. Konagaya et al. (2015) showed that spermatophore nutrition had little effect on reproductive output and clarified that the reproductive output depended on the female size rather than the mating number. In addition, although apyrene spermatozoa decreased more quickly than eupyrene spermatozoa from the spermatheca of the pipevine swallowtail females in the field (Fig. 7.3), a considerable number of apyrene spermatozoa still remained in the spermatheca even after the collapse of the spermatophore, suggesting that apyrene spermatozoa could exist in the spermatheca throughout the female lifespan (Fig. 7.4). On the other hand, Lewis and Wedell (2007) showed that the number of both eupyrene and apyrene spermatozoa decreased with consecutive mating in the males of the squinting bush brown, *Bicyclus anynana*, which females are monandrous (lifetime mating number is 1.3), because of sperm depletion with age under short longevity, despite feeding on nitrogen-rich rotting fruits. Furthermore, there seems to be a trade-off between sperm production and male longevity in monandrous species.

Boggs (1986) reported that females of the Mormon fritillary *Speyeria mormonia*, which is monandrous, fed on enough nectar, and survived for 19 days in captivity, laying eggs for 14 days, which lifespan is rather short among butterfly species. Wiklund et al. (2003) stated that male mating success of monandrous species should not be expected to be strongly associated with lifespan, because receptive females are available for a rather short time period, the exact timing of which depends on the life cycle. The mortality rate of the adult peacock butterfly, *Inachio io*, that is monandrous, is sex specific, with males having a much shorter lifespan than females. On the other hand, there is little difference in the mortality rate between sexes in most polyandrous species such as the black swallowtail butterflies (Kiritani et al. 1984), because females must be unreceptive for a number of days after mating, and so polyandrous females must distribute their mating substantially over time, during which period they may oviposit.

In the small copper, *Lycaena phlaeas*, females have few mature eggs following eclosion, and they use their own energy or own nectar intake to develop eggs without ejaculates from males (Watanabe and Nishimura 2001). The lifetime

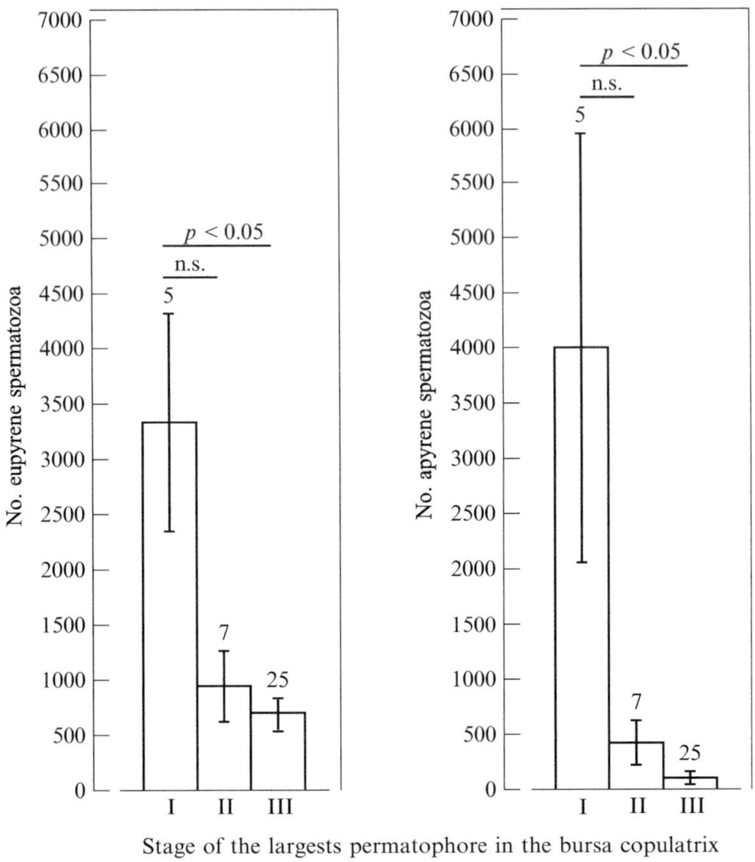

Fig. 7.3 The number of eupyrene and apyrene spermatozoa stored in the spermatheca of females in the pipevine swallowtail, *Battus philenor*, classified by the stage of the largest spermatophore (stage I–III) in the bursa copulatrix, an indicator of how recently she had mated. See Fig. 4.10 for each stage of the spermatophore. The *number* in each *bar* represents the sample size. Mann–Whitney *U* test with Bonferroni correction was used to detect changes of the number of spermatozoa from soon after the copulation (stage I). (From Konagaya et al. 2015)

reproductive output was estimated at about 150 eggs throughout their lifespan, which is 3 weeks in the laboratory. Figure 7.5 shows the change in the number of mature eggs in the females reared in the laboratory. Very few mature eggs were found in the ovaries of newly emerged females (0 day old). It follows that maturation of eggs occurs with aging. About 50 mature eggs had accumulated 1 week after emergence. Figure 7.5 also shows that the number of mature eggs in mating females is greater than the mean number of mature eggs of virgin females, and that the number of females avoiding mating is less. An increase in the number of mature eggs thus appears to facilitate female willingness to mate. Therefore, the number of mature eggs in the ovaries indicates the mating behaviour of the small copper females, indicating that there is a certain period of sexual immaturity including

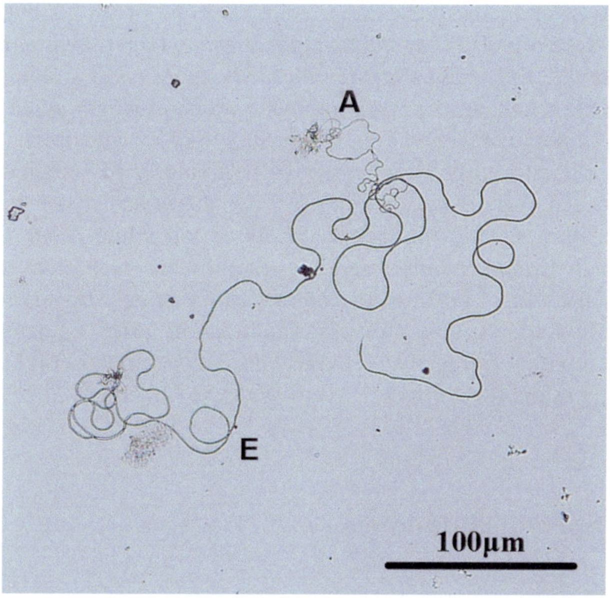

Fig. 7.4 Photograph of eupyrene spermatozoon (*E*) and apyrene spermatozoa (*A*) in the spermatheca of an old female of the pipevine swallowtail, *Battus philenor*, under the stereoscopic microscope. ×75. (From Konagaya et al. 2015)

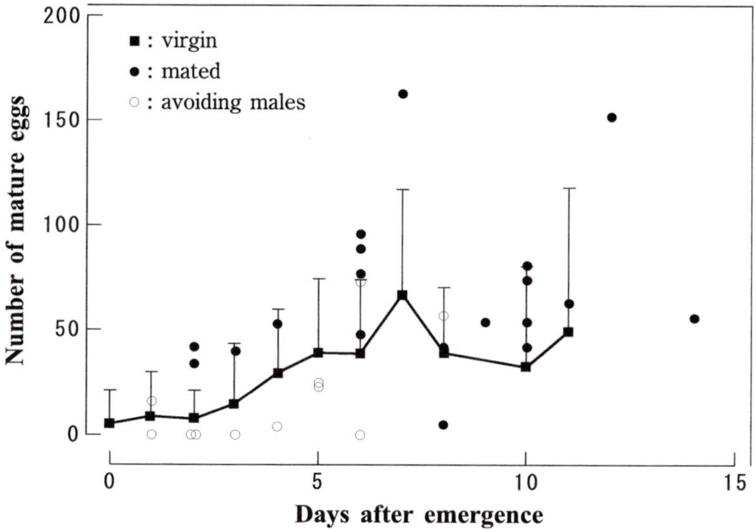

Fig. 7.5 Daily change in the number of mature eggs of virgin females of the small copper, *Lycaena phlaeas* (*squares*, +SD), and the number of mature eggs of females accepting copulation (*closed circles*) and of those avoiding males (*open circles*) in laboratory experiments (From Watanabe and Nishimura 2001)

avoidance or refusal against males after eclosion in monandrous butterfly species. Females must have a pre-mating adult stage, during which they mature their initial eggs. Consequently, this could allow females to be more particular in their choice of mate, because the small copper is a perching species, in which males spend part of the day sitting on some object waiting for passing females who are then pursued and courted. Such a territorial behaviour may be particularly advantageous in females of the monandrous species because the females could select the males. Wiklund (1977) showed that female monogamy in the wood white, *Leptidea sinapis*, is maintained by the females simply remaining quiescent during courtship, without any kind of mate refusal posture. In contrast to the small copper, females of the pipevine swallowtail, *Battus philenor*, mate soon after eclosion (Konagaya et al. 2015), notwithstanding that females show clear mate choice related to the dorsal hindwing coloration of males (Rutowski and Rajyaguru 2013).

Control of Sperm Abundance

When females mate multiply, sperm from different males is stored in the sperm storage organs, the bursa copulatrix and spermatheca. Although mate competition among males appears to end once a male has copulated with a female, such competition can be considered to extend to processes that take place during and after copulation in the female reproductive organs. The process is called sperm competition, whereby sperm from different males compete to fertilise a female's ova. Females might develop a technique to select suitable sperm in their reproductive organs. Such a cryptic female choice is also reported in many species for exerting a choice of male phenotypes (Thornhill 1983).

In polyandrous butterflies, after a female has mated multiply, most eggs laid are fertilised by the sperm of the last mated male (last male sperm precedence), as observed in the mocker swallowtail, *Papilio dardanus* (Clark and Shepard 1962), the alfalfa butterfly, *Colias eurytheme* (Boggs and Watt 1981), the green-veined white, *Pieris napi* (Bissoondath and Wiklund 1997), the cabbage white, *Pieris rapae* (Wedell and Cook 1998) and the checkerspot butterfly, *Euphydryas editha* (Labine 1966). These species exhibit bimodal P_2 values (the proportion of eggs sired by the second of two males to mate with the same female) with peaks of 0 % and 100 %, indicating that little sperm mixing has occurred. LaMunyon and Eisner (1993) suggested that the bimodal distribution of the P_2 value in the bella moth, *Utetheisa ornatrix*, is a manifestation of cryptic female choice. Consequently, females of many lepidopteran species seem to be able to 'decide' which male sperm to utilise to fertilise her eggs. Females must choose sperm based on male phenotype, such as body size, ejaculate size, age and mating history. Indeed, a number of studies on butterflies and moths have revealed that spermatophore size is critical in determination of the male's fertilisation success by the female (Bissoondath and Wiklund 1997).

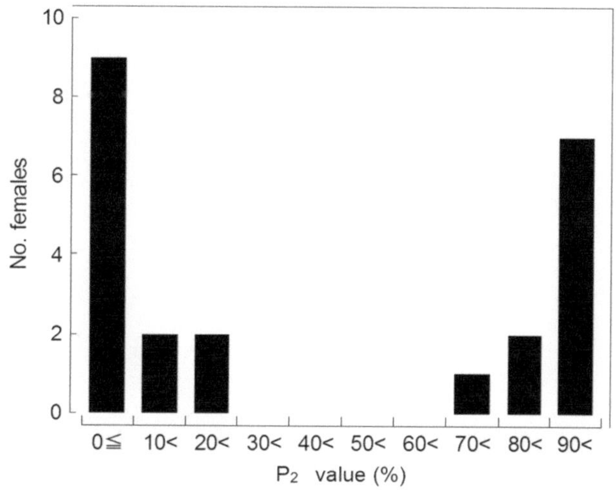

Fig. 7.6 Frequency distribution of P_2 values in twice-mated females of the swallowtail butterfly, *Papilio xuthus* (From Sasaki et al. 2015)

Watanabe and Sasaki (2010) showed that the sperm ejection must be the cause of the sperm precedence pattern in the swallowtail butterfly, *Papilio xuthus*. As shown in Fig. 7.6, it also exhibited a bimodal distribution of P_2 with peaks of 0 % and 100 % (Sasaki et al. 2015). Once sperm ejection occurs, almost all sperm of the first male in the spermatheca disappears before the migration of the sperm of the second male, and the remaining sperm are placed in the background of the spermatheca by the arrival of the second male's apyrene sperm. Consequently, most eggs laid after the second mating must be fertilised by the second male's sperm. On the other hand, when sperm ejection does not occur, little sperm of the second male would enter the spermatheca, because the spermatheca will be completely filled by the sperm of the first male. Because the spermatheca of the lepidopteran female may have a restricted storage capacity (Watanabe and Hachisuka 2005), excess sperm migration from the spermatophore may cause the spermatheca to overflow. LaMunyon (2000) reported that the storage capacity of the spermatheca of the tobacco budworm, *Heliothis virescens*, is approximately that of the mass of one ejaculate. An alternative to the first male sperm priority hypothesis is that females that had received a smaller spermatophore from the second male did not allow the second male's sperm to enter the spermatheca. Sperm migration from spermatophore to spermatheca is dependent largely on the female's musculature (Tschudi-Rein and Benz 1990), and Curril and LaMunyon (2006) suggested that females of the bella moth, *Utetheisa ornatrix*, shunt the sperm of unwanted males to different organs rather than to the spermatheca.

The possibility that the bimodal distribution of the P_2 value occurred independently of sperm ejection still remains. Harvey and Parker (2000) used mathematical simulation to demonstrate that bimodal distribution of the P_2 value can be explained by the low mixing potential of sperm. Accordingly, even in species in which the P_2 value exhibits a bimodal distribution with peaks of 0 % and 100 %, as previously mentioned in the swallowtail butterfly, *Papilio xuthus* (Sasaki

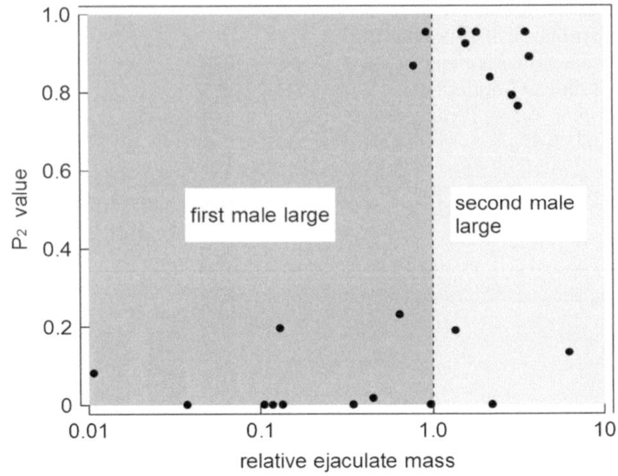

Fig. 7.7 Relationship between the relative ejaculate mass [log(weight gain of females from second mating/that from first mating)] and P_2 values in twice-mated females of the swallowtail butterfly, *Papilio xuthus* (From Sasaki et al. 2015)

et al. 2015), male fertilisation success might be affected by the relative amount of sperm in the spermatheca. Males that transfer a larger spermatophore are more likely to be principal sires (Fig. 7.7). A large spermatophore might contain much sperm (He and Miyata 1997). However, little information about sperm dynamics transferred from each male in the spermatheca has been reported. Because of the restricted capacity of the spermatheca (Watanabe and Hachisuka 2005), female participation in sperm migration (Curril and LaMunyon 2006) and so on, the number of sperm stored in the spermatheca could not be correlated with the number of sperm transferred.

Females use sperm from a male that transferred a larger spermatophore for oviposition. The benefits of producing a large spermatophore have been discussed in many lepidopteran species from the male perspective (Gwynne 2008). The spermatophore represents a male's mating effort, because a large spermatophore can increase the female refractory period (Sugawara 1979), resulting in more eggs fertilised by his sperm. The spermatophore also functions as a paternal investment, because nutrients contained in the spermatophore are used by a female to increase her longevity (Boggs and Watt 1981) and reproductive output (Watanabe 1988). As male spermatophore production ability is heritable (Wedell 2006), females must gain indirect benefit by choosing sperm from a male that transferred a larger spermatophore. However, most eggs laid until the next mating will be fertilised by only one last mated male chosen by the female. Furthermore, ejaculate is costly (Bissoondath and Wiklund 1997). Thus, males should try to mate with a female that they believe will choose their sperm.

For the female swallowtail butterfly, *Papilio xuthus*, after copulation in the laboratory, 9,000 eupyrene and 265,000 apyrene spermatozoa are transferred to a single spermatophore, and then 1,000 eupyrene and 1,100 apyrene spermatozoa successfully arrive in the spermatheca (Watanabe and Hachisuka 2005). Comparing with the lifetime number of eggs laid, an excess number of eupyrene spermatozoa is found in the spermatheca for fertilisation, if a single spermatozoon fertilises a single egg. For polyandrous lepidopterans such as the oriental armyworm, *Pseudaletia separata* (He et al. 1995), the green-veined white, *Pieris napi* (Wiklund et al. 1993), the tobacco budworm, *Heliothis virescens* (LaMunyon 2000), the bella moth, *Utetheisa ornatrix* (Curril and LaMunyon 2006) and the cotton bollworm, *Heliothis armigera* (Yan et al. 2013), an excess of eupyrene spermatozoa in the spermatheca could be common. Wild females of the monandrous pipevine swallowtail, *Battus philenor*, also retain more than 2,000 eupyrene spermatozoa (Konagaya et al. 2015).

Pre-Copula Female Choice

In males, sexual selection favours the traits that increase the number of matings and sperm competitive ability. The former is so-called precopulatory sexual selection, and the latter postcopulatory sexual selection, where mating order might have a significant role, with possible outcomes including last-male precedence. Although extended periods between matings also cause production of larger ejaculates with sperm accumulation (Vande Velde et al. 2011), there seems to be a young male mating advantage. For both sexes, senescence basically starts at the adult eclosion, indicating that older males are in poorer condition. Therefore, females might use the courtship of males to decide whether the male is a suitable age for mating. Male quality, of either genotypic or phenotypic origin, is a factor that can affect female reproductive success. For example, females mated with males of phenotypically poor quality must have significantly lower fecundity and probably lower fertility than those mated with high-quality males. Therefore, one of several reasons for rejecting a male is that he is old. As the males of most species age, they might generally produce smaller spermatophores, so that the reward for mating with an old male may simply be too low, as pointed out by Vane-Wright (2003). Wing wear and other cues may indicate the male's relative age.

In the pale clouded yellow, *Colias erate*, females show ascending flight or mate refusal posture when they do not want to copulate (Nakanishi et al. 1996). Interference from male courtship drives ovipositing females from the flight areas (Shapiro 1970). In addition, because the copulation typically lasts for about an hour, females suffer from restricted oviposition time. Thus, females have to gain benefits by avoiding harassment by males as well as prolonged copulation and excessive matings. Mate refusal posture or ascending flight in females of the pale clouded yellow often leads to interruption of the male courtship behaviour (Hasegawa and Watanabe 2008), functioning as signals of unreceptivity. However, even when the female shows the mate refusal posture, some males continue courtship behaviour, achieving copulation (Watanabe et al. 1997). The female's abdominal position must alter to control the male's behaviour (Watanabe and Irie 2011), because it must be physically impossible for the male to put his genitalia onto the tip of the female's genitalia in the hindwings for copulation (Fig. 2.4). Pierid mate refusal posture, in which unreceptive females lift their abdomen in their wings, can also avoid copulation (Suzuki 1981). On the other hand, in monandrous species such as the small copper, *Lycaena phlaeas*, wing fluttering at the perching site functions as refusal of male courting (Watanabe and Nishimura 2001).

Watanabe et al. (1997) suggested that males of the pale clouded yellow, *Colias erate*, could identify the age of males copulating, and that females may choose males based on their sex pheromones and the scales on the wings of males. The degree of pheromone secretion and worn wing condition depend on male age. Then, males can realise the receptivity of female by female abdominal posture, and they will not attempt to copulate while females hide the abdomen. Although such polyandrous *Colias* females accept re-mating depending on the intensity of the male's courtship (Rutowski 1985), prolonged courtship behaviour costs males reduced searching time for the next females as well as nectar-feeding time. Thus, interrupting courtship behaviour to an unreceptive female is adaptive for males.

Because female butterflies that copulated with a recently mated male receive a smaller spermatophore with fewer resources than in a spermatophore from a virgin male, the females must have two behavioural strategies: discriminate between males based on their mating history, and mate more than once to replenish her sperm stock. The genetic diversity of offspring and nutritional benefits might increase when the females of polyandrous species have overcome the risk of increase in time loss from male harassment. Thus, females have been shown to choose a male based on several male characteristics including size, as well as age or mating status (Rutowski 1984). In the green-veined white, *Pieris napi*, females first mated to small-size males tend to re-mate sooner than females first mated to large-size males (Bissoondath and Wiklund 1997). The large-size male might have a selective advantage over small-size males, because both the male fertilisation success and the female refractory period are influenced by the ejaculate size transferred. On the other hand, Vande Velde et al. (2011) suggested that females are able to detect the cues of males from previous mates and use this information to avoid mating with sperm-exhausted males.

Sexual selection is expected to favour traits increasing the number of matings and sperm competitive ability, resulting in increased fecundity or lifespan. Then, females are apt to mate with males in a way that maximises their individual fitness, whereas male mating success is positively related to proxies, indicating resource-holding potential such as body mass or fat body content. In the case of the butterfly, because adults no longer grow after eclosion, physiological and morphological conditions including body mass, fat content and the degree of worn wing diminish with age. Additionally, spermatophore mass, sperm number and sperm quality must be expected to decrease with male age. Therefore, females seem to mate preferentially with younger rather than older males. However, there is an old male mating advantage. Karl and Fischer (2013) classified the outcome of advantage into three for the squinting bush brown, *Bicyclus anynana*, as follows:

1. Increased competitive ability or eagerness to mate in older compared to younger males
2. Females intrinsically preferring older over younger males
3. A combination of both factors

The results showed that females that mated with older compared to younger males exhibited reduced hatching success of eggs fertilised by the sperm of older males, a higher re-mating propensity and shorter refractory periods for the subsequent mating. Thus, mating with younger males is advantageous for reproductive success from the female perspective.

The higher re-mating frequency of females that had mated initially with older compared to younger males must indicate that the old male mating advantage is not the result of female mate preference. Females seem to compensate for reproductive disadvantages by mating more readily with additional males. Therefore, older males have to acquire a mating advantage by being more aggressive and persistent during courtship, as in the pale clouded yellow, *Colias erate* (Watanabe et al. 1997).

Post-Copula Female Choice

Postcopulatory sexual selection is potentially important for males against polyandrous females, whereas reproductive success of the male does not necessary increase with the numbers of his copulations. Males have to compete for fertilisation of the eggs even after copulation, that is, sperm competition, which is competition among sperm of different males to fertilise the eggs of the female (Birkhead and Møller 1998). Therefore, sperm competition might be of major importance in the evolution of reproductive strategies and mating systems. On the other hand, female-imposed mechanisms that bias paternity towards males with preferred traits over other males occur when the female has mated multiply. Cryptic female choice has been proposed. Thereafter, substantial indirect evidence of the cryptic female choice as a relationship between copulatory courtship intensity and sperm precedence has been reported in numerous taxa (Eberhard 1991). However, few studies have conclusively showed that paternity bias is driven by cryptic female choice (Fedina 2007).

One of the best known possible mechanisms of the cryptic female choice is internal sperm manipulation. Female must determine which males sire their young even after copulation has occurred by prematurely terminating copulations, failing to store transferred sperm, removing or ejecting stored sperm and other tactics. Eberhard (1996) stated that to show the particular female process or traits in the cryptic female choice mechanism, it is necessary to demonstrate the following:

1. Female responses to some conspecific males differ from those to others
2. Discrimination among males results in differences in reproductive success for the particular males
3. Female biases are associated with particular male characteristics

Thus, many potential mechanisms of cryptic female choice such as premature interruption of copulation, lack of sperm transport to the storage organ, ejection of sperm from the storage organ, lack of ovulation and selective abortion have been proposed.

Of these mechanisms proposed by Eberhard (1996), disappearance of sperm in the female sperm storage organ, the spermatheca, has been clarified in several species of Lepidoptera (Pair et al. 1977). Sperm in the spermatophore start to migrate to the spermatheca several hours after the termination of copulation (Watanabe and Hachisuka 2005). After migration, sperm are stored in the spermatheca and used for fertilisation. Because there is an excess number of sperm in the spermatheca compared with the number of eggs needing fertilising in the female, it has been believed that at least a certain number of sperm in the spermatheca survive throughout the female's lifespan. However, Etman and Hooper (1979) showed that in the oriental leafworm, *Spodoptera litura*, sperm in the spermatheca of re-mated females started to decrease and the spermatheca became empty just after the termination of copulation and then increased again. The morphological restrictions of the aedeagus organs of lepidopteran males prevent the males from removing rival sperm mechanically from the spermatheca by using their aedeagus organs. Lepidopteran males are only able to insert their accessory organs up to the entrance of the bursa copulatrix. Consequently, males cannot gain access to the spermatheca. In addition, because the disappearance of sperm in the spermatheca occurs before the sperm migration of the current male has started (Xu and Wang 2010), sperm in the spermatheca are not washed away by the current male's ejaculate. Thus, females, not males, must control this process. Watanabe and Sasaki (2010) demonstrated that sperm disappearance in the swallowtail butterfly, *Papilio xuthus*, was more likely to occur when females re-mated with males that transferred a large spermatophore, suggesting that sperm ejection is the mechanism by which paternity is biased towards preferred males. Therefore, lepidopteran females must have the ability to determine whether to eject sperm from a former mated male on the basis of the traits of her mates.

Spermatophore size might be used for a cue for cryptic female choice. LaMunyon and Eisner (1994) stated that rather than utilising the sperm of all her mates in polyandrous species, the female uses predominantly those from larger partners that provide more rich nutrients and pyrrolizidine alkaloid in the larger spermatophore. The former enables the female to increase her fecundity, and the latter is bestowed upon the eggs by her to protecting against predation.

Watanabe and Sasaki (2010) clarified that sperm ejection by female of the swallowtail butterfly, *Papilio xuthus*, occurred when the females had received a larger spermatophore from the second male. At 3 h after the termination of copulation, no sperm had arrived in the spermatheca (Fig. 7.8). A few apyrene spermatozoa were found in the spermatheca 8 h after the termination of copulation, and they were still active. Some apyrene sperm mass actively moving in the reproductive duct to the spermatheca (mainly in the spermathecal duct) was observed. Eupyrene spermatozoa began to arrive 12 h after the termination of copulation, during which apyrene spermatozoa had filled the spermatheca and were still moving actively. Each spermatophore lacked both types of spermatozoa at 24 h after copulation, suggesting that both eupyrene and apyrene sperm migration had ended within 24 h after the termination of copulation. At 3 days and 7 days after copulation, during which the mated females were allowed to oviposit, the

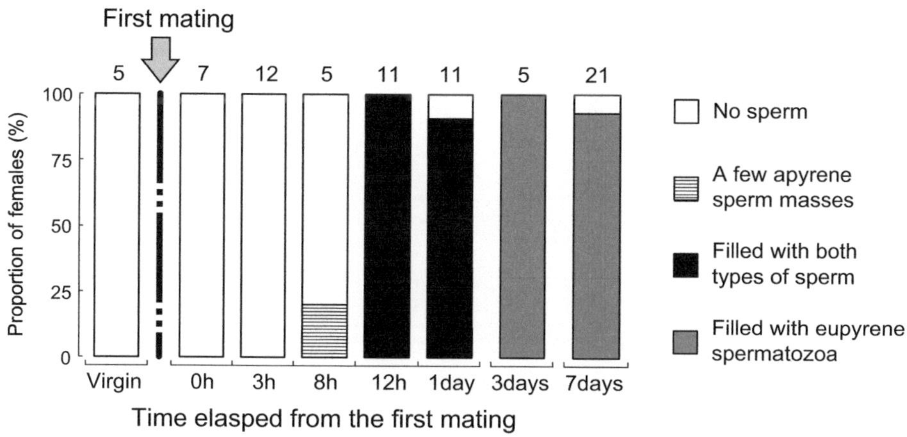

Fig. 7.8 Changes in the percentage of the females of the swallowtail butterfly, *Papilio xuthus*, with spermatheca containing sperm after the first mating. *Numerals* indicate number of females dissected (From Watanabe and Sasaki 2010)

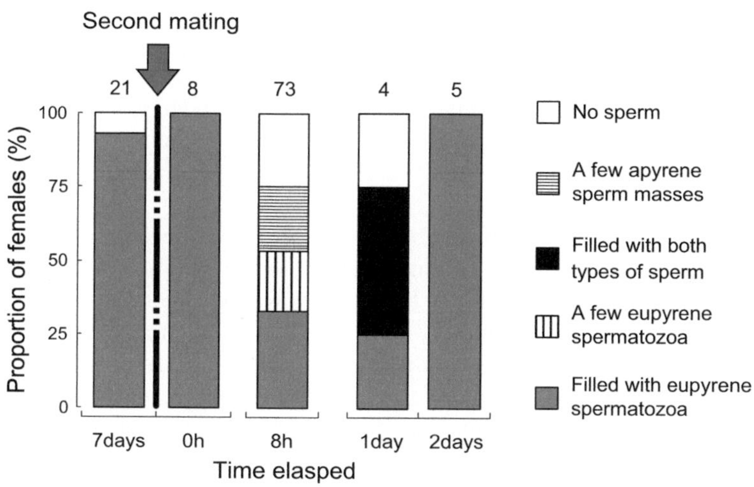

Fig. 7.9 Changes in the percentage of the females of the swallowtail butterfly, *Papilio xuthus*, in relationship to sperm occupying the spermatheca after the second mating. *Numerals* indicate number of females dissected (From Watanabe and Sasaki 2010)

spermatheca contained many eupyrene spermatozoa and a few apyrene spermatozoa, both nonactive. In spite of oviposition activity, there was no change in the shape and volume of the spermatheca, although the spermatophore was collapsed.

On the other hand (Fig. 7.9), re-mated females had two spermatophores in the bursa copulatrix. One was small and collapsed without any spermatozoa, which was identified to be derived from the first male. The other was intact, teardrop shaped,

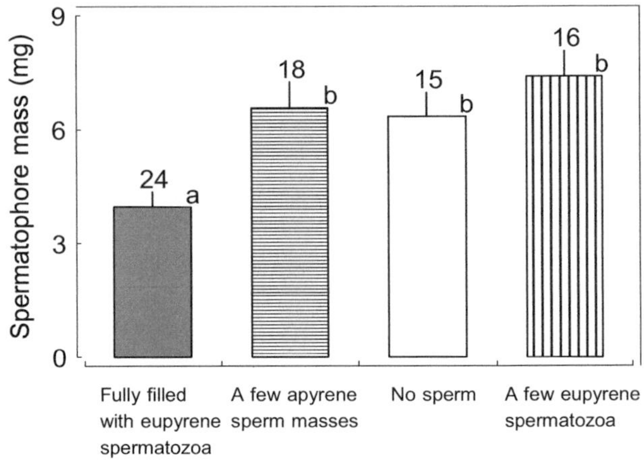

Fig. 7.10 Spermatophore mass of the second male in relationship to the quantity and quality of sperm in the spermatheca of the re-mated females of the swallowtail butterfly, *Papilio xuthus*, 8 h after the termination of the copulation (+SE). Different letter indicates a significant difference at probability (P) <0.05, using Dunnett T_3 test. *Numerals* indicate number of females examined (From Watanabe and Sasaki 2010)

with many eupyrene sperm bundles and many active apyrene sperm masses, and usually large, indicating that it was from the second male. Spermatheca of the re-mated females dissected just after the termination of copulation were filled by many eupyrene spermatozoa and a few nonactive apyrene spermatozoa, both of which were derived from the first male. Then, 18 of 73 re-mated females dissected 8 h after the termination of the copulation had a vacant spermatheca. The proportion of females with the spermatheca filled by sperm was significantly lower than that of the females just before re-mating. Thereafter, the number of females without sperm in the spermatheca decreased, and all re-mated females had sperm 48 h after the termination of the copulation.

The spermatophore mass in the bursa copulatrix of the second mating was examined in relationship to the sperm condition in the spermatheca 8 h after the termination of copulation (Fig. 7.10). For females with the spermatheca fully filled by sperm derived from the first male, the second spermatophore weight was 4.0 mg, which was significantly smaller than that of the first mating. On the other hand, the second spermatophore weight in the females with the spermatheca including a few or no sperm was more than 6 mg, the normal size. Both were significantly larger than that of the second spermatophore with fully filled sperm of the first male, respectively. The spermatophore weight in females with spermatheca filled by the sperm of the second male was also 6 mg, which is larger than that with the first male. Therefore, the spermatophore mass might affect the sperm condition in the spermatheca, and sperm ejection might be the mechanism that biases paternity towards preferred males that had transferred larger ejaculate and gained higher paternity. Accordingly, the bimodal distribution of P_2 values is confirmed to have been achieved by sperm ejection by the females.

Females of some lepidopteran species could exercise the cryptic female choice by sperm ejection. Etman and Hooper (1979) showed that, in the oriental leafworm moth, *Spodoptera litura*, the spermatheca of re-mated females became empty just after the termination of copulation and then refilled. Pair et al. (1977) reported that, in tobacco budworm females of the hybrid between *Heliothis subflexa* and *Heliothis virescens*, the eupyrene spermatozoa in the spermatheca were decreased by 88 % after re-mating. Simmons (2001) pointed out that a bimodal distribution of P_2 value is common in Lepidoptera. Xu and Wang (2010) observed sperm ejection from female spermatheca in the Mediterranean flour moth, *Ephestia kuehniella*, which exhibited last male sperm precedence, indicating that sperm ejection by a female results in not only a bimodal distribution but also last male sperm priority. Therefore, sperm ejection might be a common mate choice mechanism for lepidopteran females.

Sperm ejection by females has been reported in several insect species. Females of the common fruit fly, *Drosophila melanogaster*, release stored sperm from the reproductive tract after copulation with a second male (Snook and Hosken 2004). The number of sperm stored in the spermatheca of damselfly females decreases before oviposition as the result of ejection by the female (Córdoba-Aguilar 2006).

There is insufficient evidence regarding how females compare the ejaculate size of the first and the second males, although females seem to be able to measure the size of ejaculate using stretch receptors in the wall of the bursa copulatrix (Sugawara 1981). For the swallowtail butterfly, *Papilio xuthus*, the size of the spermatophore was significantly decreased 3 days after mating (Watanabe et al. 2000). Because the time interval of the first and second matings was 3 to 4 days, the size of the ejaculate from the first male just after the second mating must be decreased. If females measure and compare the size of both ejaculates just after the second mating, the sperm of the second male might always be chosen when the female received ejaculates of the same size. However, the first male's sperm was used by a female when the first male's ejaculate had been larger than that of the second male, suggesting that *P. xuthus* females choose sperm based on the size of both spermatophores just after transferring.

Mating Strategy of Overwintering Females

Where there are two or more generations of butterfly species each year, it is usually the last generation which hibernates. In temperate regions, because winter is an unsuitable season to be active because of low temperatures and intermittent snow, mortality must increase greatly during overwintering, and many butterfly species remain in inactive stages such as eggs or pupae to hibernate, mainly under a hormonally controlled diapause. Host plants for larvae are also not widely available during the winter season in temperate zones, nor are nectar plants available for adults. Consequently, any reproductive behaviour of adult butterflies during the winter is not adaptive. However, some species of butterflies such as the comma butterfly, *Polygonia c-album*, overwinter in the adult stage with reproductive diapause, involving the arrest of reproductive organ by hormonal control (Nylin 1992). They seem to live much longer by gradually using their fat body reserves, because of their slow metabolic rate, mainly dwelling in the forests that might offer protection against environmental extremes rather than in the open grasslands. Then, adult butterflies become active again at the onset of spring. Most of them are receptive for copulation after hibernation, although they have already mated before hibernation. Mated females of the monarch butterfly, *Danaus plexippus*, use nutrients in the spermatophore during the overwintering period (Leong et al. 2012). On the other hand, nuptial gifts to overwintering females are believed to be also beneficial for males with which they had mated before overwintering because the "overwintering success of sperm" depends on the mortality of females.

The common grass yellow, *Eurema mandarina*, is a butterfly that shows seasonal polyphenism in wing colour consisting of summer-morphs and autumn-morphs. Low temperature and short photoperiod during the larval stage induce autumn-morph adults, and sequential low temperatures lead them to overwinter with reproductive diapause (Yata 1974). Only autumn-morph adults are observed the following spring. Because of the presence of overlapping generations in late autumn and the difference in critical photoperiod between males and females (Kato and Sano 1987), summer-morph males but not females fly sympatrically with both autumn-morph males and females in October to November. Although autumn-morph males show little mating activity in the autumn (Kato 1989), mating between summer-morph males and autumn-morph females has frequently been observed. Then, most autumn-morph females before overwintering already contain a single spermatophore in their bursa copulatrix, and autumn-morph females re-mate with autumn-morph males the following spring. Thus, mating of the autumn-morph females between two physiologically different generations commonly occurs in this species.

Two hypotheses for the adaptive significance of the multiple mating of females of the common grass yellow, *Eurema mandarina*, with both summer- and autumn-morph males have been proposed: the nuptial gift hypothesis and the risk-hedge hypothesis (Kato 1986). Although autumn-morph females are sometimes active at sunflecks in deciduous forests during the winter, they must not obtain nutrients or energy because of the absence of flowering plants. Thus, the spermatophore derived from the summer morph males might be an important nutrient for the females during overwintering. The females might delay ovarian development during the overwintering because low temperatures affect metabolic rate. In addition, there is a risk for females that, after overwintering, they may encounter no males to mate with until the sprouting of the host plants because of spring weather conditions or biological factors. On the other hand, mated females are better able to handle the physiological stresses of rapid egg maturation when they re-mate with the autumn-morph males after overwintering (Leong et al. 2012). Thus, it might be adaptive for such females to copulate with the summer-morph males and lay eggs inseminated by their sperm.

The long-term storage of viable sperm in the post-overwintering females has been reported on the common grass yellow, *Eurema mandarina* (Konagaya and Watanabe 2015) and the monarch butterfly, *Danaus plexippus* (Leong et al. 2012). The physiology of the storage and the maintenance of viable sperm in the spermatheca of females during overwintering might be the result of the seminal fluids from the males mated before overwinter, the spermathecal gland secretions of the females themselves and both combined.

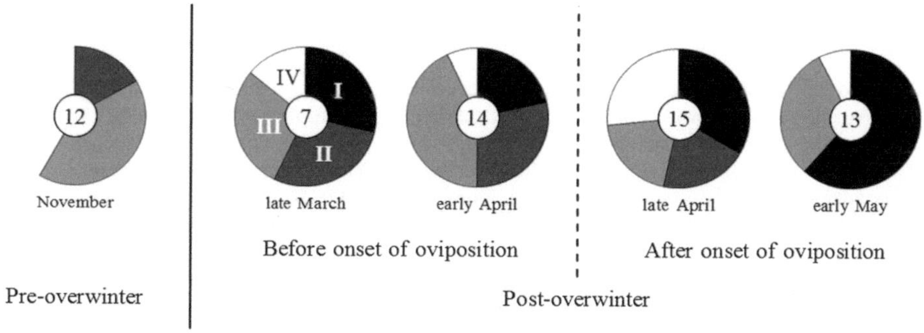

Fig. 7.11 Females of the common grass yellow, *Eurema mandarina*, classified by the stage of the largest spermatophore in the bursa copulatrix for each survey period. The onset of oviposition occurred in mid-April after overwintering. Of 12 females captured in November, 7 had mated and 5 were virgins. The shape and the condition of spermatophores in each stage are shown in Fig. 4.10 (From Konagaya and Watanabe 2015)

In the common grass yellow, *Eurema mandarina*, the timing of the re-mating with autumn-morph males and the onset of oviposition in autumn-morph females is important for summer-morph males that had mated before winter (Konagaya and Watanabe 2015). When the autumn-morph females re-mate with the autumn-morph males before the onset of oviposition, sperm derived from the summer-morph males must be exposed to severe sperm competition. Thus, when autumn-morph females re-mate with autumn-morph males before the onset of oviposition, summer-morph males lose their advantage. The autumn-morph females start oviposition in mid-April. Change in the shape of the spermatophore in the wild females indicates that most autumn-morph females re-mate with autumn-morph males before mid-April, the onset of oviposition (Fig. 7.11). Although the eupyrene sperm of the summer-morph males remain in the spermatheca after overwintering, only a few autumn-morph females lay eggs with sperm inseminated by the summer-morph males. Therefore, the nuptial gift hypothesis is supported.

After overwintering, the autumn-morph males of the common grass yellow, *Eurema mandarina*, increase movement activity, as if they were trying to search for autumn-morph females (Kato 1989). Consequently, the females may have many opportunities to encounter autumn-morph males, making the risk-hedge hypothesis unlikely. Although males attempt to reduce female receptivity to re-mating through the physical stimulus of ejaculating a large spermatophore, on the other hand, the spermatophore derived from summer-morph males was observed to have declined in the bursa copulatrix of females after overwintering, suggesting that it is difficult for summer-morph males to prevent their mates from re-mating after overwintering using spermatophore size. Konagaya and Watanabe (2015) suggested that the larger eupyrene spermatozoon of summer-morph males in comparison to that of autumn-morph males might be a counter-adaptation to the mating strategy of autumn-morph females.

References

Birkhead TR, Møller AP (1998) Sperm competition and sexual selection. Academic Press, London

Bissoondath CJ, Wiklund C (1995) Protein content of spermatophores in relation to monandry/polyandry in butterflies. Behav Ecol Sociobiol 37:365–371

Bissoondath CJ, Wiklund C (1997) Effect of male body size on sperm precedence in the polyandrous butterfly *Pieris napi* L. (Lepidoptera: Pieridae). Behav Ecol 8:518–523

Boggs CL (1986) Reproductive strategies of female butterflies: variation in and constraints on fecundity. Ecol Entomol 11:7–15

Boggs CL, Watt WB (1981) Population structure of pierid butterflies. IV. Genetic and physiological investment in offspring by male *Colias*. Oecologia (Berl) 50:320–324

Clark CA, Shepard PM (1962) Offspring from double matings in swallowtail butterflies. Entomologist 95:199–203

Córdoba-Aguilar A (2006) Sperm ejection as a possible cryptic female choice mechanism in Odonata (Insecta). Physiol Entomol 31:146–153

Curril IM, LaMunyon CW (2006) Sperm storage and arrangement within females of the arctiid moth *Utetheisa ornatrix*. J Insect Physiol 52:1182–1188

Eberhard WG (1991) Copulatory courtship and cryptic female choice in insects. Biol Rev 66:1–31

Eberhard WG (1996) Female control: sexual selection by cryptic female choice. Princeton University Press, Princeton

Elzinga JA, Chevasco V, Grapputo A, Mappes J (2011) Influence of male mating history on female reproductive success among monandrous Naryciinae (Lepidoptera: Psychidae). Ecol Entomol 36:170–180

Etman AAM, Hooper GHS (1979) Sperm precedence of the last mating in *Spodoptera litura*. Ann Entomol Soc Am 72:119–120

Fedina TY (2007) Cryptic female choice during spermatophore transfer in *Tribolium castaneum* (Coleoptera: Tenebrionidae). J Insect Physiol 53:93–98

Gwynne DT (2008) Sexual conflict over nuptial gifts in insects. Annu Rev Entomol 53:83–101

Harvey IF, Parker GA (2000) 'Sloppy' sperm mixing and intraspecific variation in sperm precedence (P_2) patterns. Proc R Soc Lond B 267:2537–2542

Hasegawa M, Watanabe M (2008) Changes in mate preference of male *Colias erate poliographus* in relation to copulation experience of females. Jpn J Entomol 11:105–114 (in Japanese with English summary)

He Y, Miyata T (1997) Variations in sperm number in relation to larval crowding and spermatophore size in the armyworm, *Pseudaletia separata*. Ecol Entomol 22:41–46

He Y, Tanaka T, Miyata T (1995) Eupyrene and apyrene sperm and their numerical fluctuations inside the female reproductive tract of the armyworm, *Pseudaletia separata*. J Insect Physiol 41:689–694

Karl I, Fischer K (2013) Old male mating advantage results from sexual conflict in a butterfly. Anim Behav 85:143–149

Kato Y (1986) The prediapause copulation and its significance in the butterfly *Eurema hecabe*. J Ethol 4:81–90

Kato Y (1989) Differences in reproductive behavior among seasonal wing morphs of the butterfly *Eurema hecabe*. J Insect Behav 2:419–429

Kato Y, Sano M (1987) Role of photoperiod and temperature in seasonal morph determination of the butterfly *Eurema hecabe*. Physiol Entomol 12:417–423

Kiritani K, Nozato K, Miyai S, Ban Y (1984) Studies on ecology and behavior of Japanese black swallowtail butterflies. I. Ecological characteristics of male populations in *Papilio helenus nicconicolens* Butler and *P. protenor demetrius* Cramer (Lepidoptera: Papilionidae). Appl Entomol Zool 19:509–517

Konagaya T, Watanabe M (2015) Adaptive significance of the mating of autumn-morph females with non-overwintering summer-morph males in the Japanese common grass yellow, *Eurema mandarinia* (Lepidoptera: Pieridae). Appl Entomol Zool 50:41–47

Konagaya T, Mutoh N, Suzuki M, Rutowski RL, Watanabe M (2015) Estimates of female lifetime fecundity and changes in the number and types of sperm stored with age and time since mating in the monandrous swallowtail butterfly, *Battus philenor* (Lepidoptera: Papilionidae) in the Arizona desert. Appl Entomol Zool 50:311–316

Labine PA (1966) The population biology of the butterfly, *Euphydryas editha*. IV. Sperm precedence—a preliminary report. Evolution 20:580–586

LaMunyon CW (2000) Sperm storage by females of the polyandrous noctuid moth *Heliothis virescens*. Anim Behav 59:395–402

LaMunyon CW, Eisner T (1993) Postcopulatory sexual selection in an arctiid moth (*Utetheisa ornatrix*). Proc Natl Acad Sci USA 90:4689–4692

LaMunyon CW, Eisner T (1994) Spermatophore size as determinant of paternity in an arctiid moth (*Utetheisa ornatrix*). Proc Natl Acad Sci USA 91:7081–7084

Leong KHL, Yoshimura MA, Williams C (2012) Adaptive significance of previously mated monarch butterfly females (*Danaus plexippus* (Linneaus)) overwintering at a California winter site. J Lepid Soc 66:205–210

Lewis Z, Wedell N (2007) Effect of adult feeding on male mating behaviour in the butterfly, *Bicyclus anynana* (Lepidoptera: Ntmphalidae). J Insect Behav 20:201–213

McNamara KB, Elgar MA, Jones TM (2009) Large spermatophores reduce female receptivity and increase male paternity success in the almond moth, *Cadra cautella*. Anim Behav 77:931–936

Nakanishi Y, Watanabe M, Ito T (1996) Differences in lifetime reproductive output and mating frequency of two female morphs of the sulfur butterfly, *Colias erate* (Lepidoptera: Pieridae). J Res Lepid 35:1–8

Nylin S (1992) Seasonal plasticity in life history traits: growth and development in *Polygonia c-album* (Lepidoptera: Nymphalidae). Biol J Linn Soc 47:301–323

Oberhauser KS (1992) Rate of ejaculate breakdown and intermating intervals in monarch butterflies. Behav Ecol Sociobiol 31:367–373

Oberhauser KS (1997) Fecundity, lifespan and egg mass in butterflies: effects of male-derived nutrients and female size. Funct Ecol 44:166–175

Odendaal FJ (1989) Mature egg number influences the behavior of female *Battus philenor* butterflies. J Insect Behav 2:15–25

Pair SD, Laster ML, Martin DF (1977) Hybrid sterility of the tobacco budworm: effects of alternate sterile and normal matings on fecundity and fertility. Ann Entomol Soc Am 70:952–954

Rutowski RL (1984) Sexual selection and the evolution of butterfly mating behavior. J Res Lepid 23:125–142

Rutowski RL (1985) Evidence for mate choice in a sulphur butterfly (*Colias eurytheme*). Z Tierpsychol 70:103–114

Rutowski RL, Rajyaguru PK (2013) Male-specific iridescent coloration in the pipevine swallowtail (*Battus philenor*) is used in mate choice by females but not sexual discrimination by males. J Insect Behav 26:200–211

Sasaki N, Todoriki S, Watanabe M (2015) Effect of sperm ejection by females on male fertilization success in the swallowtail butterfly, *Papilio xuthus* L. (Lepidoptera: Papilionidae). J Res Lepid 48:39–44

Shapiro AM (1970) The role of sexual behavior in density-related dispersal of pierid butterflies. Am Nat 104:367–372

Simmons LW (2001) Sperm competition and its evolutionary consequences in the insects. Princeton University Press, Oxford

Snook RR, Hosken DJ (2004) Sperm death and dumping in *Drosophila*. Nature (Lond) 428:939–941

Sugawara T (1979) Stretch reception in the bursa copulatrix of the butterfly, *Pieris rapae crucivora*, and its role in behaviour. J Comp Physiol 130:191–199

Sugawara T (1981) Fine structure of the stretch receptor in the bursa copulatrix of the butterfly, *Pieris rapae crucivora*. Biomed Life Sci 217:23–36

Suzuki Y (1981) Repertories of the mate avoidance behavior in the cabbage butterfly. Chou-Chou 4(8):27–29 (in Japanese)

Svärd L, Wiklund C (1989) Mass and production rate of ejaculates in relation to monandry/polyandry in butterflies. Behav Ecol Sociobiol 24:395–402

Thornhill R (1983) Cryptic female choice and its implications in the scorpionfly *Harpobittacus nigriceps*. Am Nat 122:765–788

Tschudi-Rein K, Benz G (1990) Mechanisms of sperm transfer in female *Pieris brassicae* (Lepidoptera: Pieridae). Ann Entomol Soc Am 83:1158–1164

Vande Velde L, Damiens D, Van Dyck H (2011) Spermatophore and sperm allocation in males of the monandrous butterfly *Pararge aegeria*: the female's perspective. Ethology 117:645–654

Vane-Wright D (2003) Butterflies. The Natural History Museum, London

Watanabe M (1988) Multiple matings increase the fecundity of the yellow swallowtail butterfly, *Papilio xuthus* L., in summer generations. J Insect Behav 1:17–29

Watanabe M, Hachisuka A (2005) Dynamics of eupyrene and apyrene sperm storage in ovipositing females of the swallowtail butterfly *Papilio xuthus* (Lepidoptera: Papilionidae). Entomol Sci 8:65–71

Watanabe M, Irie S (2011) Female abdomen position signals receptivity during courtship in the Japanese sulphur butterfly, *Colias erate* (Pieridae). J Lepid Soc 65:259–263

Watanabe M, Kobayashi T (2006) Total sperm ejaculation in monandrous (*Papilio machaon*) and polyandrous (*P. xuthus*) swallowtail butterflies (Lepidoptera: Papilionidae) restricted to larval stage-derived nutrients. J Res Lepid 39:1–7

Watanabe M, Nishimura M (2001) Reproductive output and egg maturation in relation to mate-avoidance in monandrous females of the small copper, *Lycaena phlaeas* (Lycaenidae). J Lepid Soc 54:83–87

Watanabe M, Nozato K (1986) Fecundity of the yellow swallowtail butterflies, *Papilio xuthus* and *P. machaon hippocrates*, in a wild environment. Zool Sci 3:509–516

Watanabe M, Sasaki N (2010) Pattern of sperm storage and migration in the reproductive tract of the swallowtail butterfly *Papilio xuthus*: cryptic female choice after second mating. Physiol Entomol 35:328–333

Watanabe M, Nakanishi Y, Bon'no M (1997) Prolonged copulation and spermatophore size ejaculated in the sulfur butterfly, *Colias erate* (Lepidoptera: Pieridae) under selective harassments of mated pairs by conspecific lone males. J Ethol 15:45–54

Watanabe M, Bon'no M, Hachisuka A (2000) Eupyrene sperm migrates to spermatheca after apyrene sperm in the swallowtail butterfly, *Papilio xuthus* L. (Lepidoptera: Papilionidae). J Ethol 18:91–99

Wedell N (2006) Male genotype affects female fitness in a paternally investing species. Evolution 60:1638–1645

Wedell N, Cook PA (1998) Determinants of paternity in a butterfly. Proc R Soc Lond B 265:625–630

Wedell N, Cook PA (1999) Butterflies tailor their ejaculate in response to sperm competition risk and intensity. Proc R Soc Lond B 266:1033–1039

Wiklund C (1977) Courtship behavior in relation to female monogamy in *Leptidea sinapis* (Lepidoptera). Oikos 29:275–283

Wiklund C, Kaitala A, Lindfors V, Abenius J (1993) Polyandry and its effect on female reproductive output in the green-veined white butterfly (*Pieris napi* L.). Behav Ecol Sociobiol 33:25–33

Wiklund C, Gotthard K, Nylin S (2003) Mating system and the evolution of sex-specific mortality rates in two nymphalid butterflies. Proc R Soc Lond B 270:1823–1828

Xu J, Wang Q (2010) Mechanisms of last male precedence in a moth: sperm displacement at ejaculation and storage sites. Behav Ecol 21:714–721

Yan S, Li H, Zhang J, Zhu J, Zhang Q, Liu X (2013) Sperm storage and sperm competition in the *Helicoverpa armigera* (Lepidoptera: Noctuidae). J Econ Entomol 106:708–715

Yata O (1974) Studies on seasonal forms and imaginal diapause in two *Eurema* species in Japan: Lepidoptera: Pieridae. Tyo to GA 25:47–54 (in Japanese with an English abstract)

Chapter 8
Apyrene Sperm as a Key Factor for Sperm Competition

Keywords Activation • Antagonistic coevolution • Blocking hypothesis • Cheap filler hypothesis • Elimination hypothesis • Flushing • Motility • Paternal investment • Position effect • Sexual conflict

Both types of sperm, eupyrene and apyrene, are individually structured in the testis as bundles. Until accumulation in the seminal vesicles, apyrene sperm bundles disappear, and free apyrene spermatozoa are ready to transfer although the eupyrene sperm is still bundled. During copulation, apyrene spermatozoa and eupyrene sperm bundles are transferred, and consequently, immediately after copulation, the spermatophore contains free apyrene spermatozoa and eupyrene sperm bundles. The eupyrene sperm bundles disappear in the spermatophore after copulation, and free apyrene spermatozoa start to migrate towards the spermatheca before the free eupyrene spermatozoa move. Although free apyrene spermatozoa arrive at the spermatheca earlier than free eupyrene spermatozoa, the spermatheca contains both apyrene and eupyrene spermatozoa. The spermathecal gland, which is a cul-de-sac tube connected to the spermatheca, is also filled gradually by apyrene and then eupyrene spermatozoa after sperm migration. Eupyrene spermatozoa in the spermatheca are used in fertilisation during oviposition, and decrease with the number of eggs laid. Changes in the number of apyrene spermatozoa both in the spermatophore ejaculated and in the spermatheca immediately after sperm migration suggest that females could absorb apyrene spermatozoa, probably for intake of additional nutrients. Many hypotheses on the role of apyrene sperm have been proposed from the aspects of developmental biology, physiology and behavioural ecology. Although the physiological aim of apyrene spermatozoa in male reproductive organs is considered to be sperm transfer, each hypothesis is focussed on sperm competition because a considerable number of apyrene spermatozoa arrive at the final destination of the female, the spermatheca, and because females receive much more apyrene spermatozoa with a single copulation, and furthermore because of counter-adaptation on the multiple mating of females. The apyrene sperm must have an important role for butterflies to consider sperm competition.

M. Watanabe, *Sperm Competition in Butterflies*, Ecological Research Monographs,
DOI 10.1007/978-4-431-55945-0_8

Theoretical Framework

Males are unlikely to have been selected to provide paternal investment before insemination because of the uncertainty of parentage resulting from internal fertilisation, including differences in the morphology of the reproductive organs in both sexes. Although a female is always certain to be the parent of the offspring she produces, a male can never be certain that he will father the female's offspring that stand to benefit from his parental investment. Furthermore, a female egg size is tens of thousands of times larger than that of a male eupyrene spermatozoon. The number of gametes produced by each sex seems to increase the inequality in parental investment. For example, a female of the swallowtail butterfly, *Papilio xuthus*, typically produces several hundred eggs in her lifetime, and fewer than 500 eggs are actually laid. On the other hand, a male can ejaculate about 20,000 eupyrene and 200,000 apyrene spermatozoa in just one copulation (Watanabe and Kobayashi 2006). Such numerous apyrene sperm have been thought to have evolved in a sperm competition context in response to the potential costs of gamete production.

Wedell et al. (2009) mentioned that sexual conflict occurs in sexually reproducing organisms and promotes rapid antagonistic coevolution of male and female reproductive traits, involving the conflict over female matings in promiscuous species, wherein males manipulate female receptivity and females resist the manipulations. In the polyandrous green-veined white, *Pieris napi*, in Sweden, a sexual conflict over female mating numbers in relationship to male nutrient provisioning has been shown. In addition, the sexual conflict over female receptivity in this species involves production and storage of apyrene spermatozoa and may be responsible for the ejaculate consisting predominantly of apyrene spermatozoa. In other words, the mating conflict involves male manipulation (apyrene sperm transfer) and female resistance (apyrene sperm storage).

In the butterfly, sperm migration from the spermatophore to the spermatheca after copulation is needed for fertilisation of the eggs. Because the eupyrene spermatozoon is the normal spermatozoon that can fertilise eggs, the paternity of the eggs will be affected by the number of eupyrene spermatozoa from each male when sperm competition occurs in the spermatheca. In addition, the apyrene spermatozoa also migrate to the spermatheca, although they do not fertilise. There are many hypotheses on the roles of the non-fertile sperm produced by males, based on structural observations or biochemical analysis as well as on sperm competition, although both the supporting role for eupyrene spermatozoa and the nutrient source for eupyrene spermatozoa or the female herself have been the principal focus. Among the several feasible hypotheses concerning the function of the apyrene spermatozoa, Friedländer (1997) reviewed these as follows:

1. The apyrene spermatozoa provide nutrients for the eupyrene spermatozoa within the female reproductive organs
2. The DNA of the discarded nuclei in the apyrene spermatozoa is metabolised and serves as a substrate for glycogen biosynthesis

3. The apyrene spermatozoa facilitate acquisition of motility by the eupyrene spermatozoa within the female reproductive organs
4. The apyrene spermatozoa have a role in the competition among the spermatozoa of the different inseminating males within the female reproductive organs
5. The apyrene spermatozoa help in transporting the eupyrene spermatozoa within the female genital duct
6. The apyrene spermatozoa make way for the migration of the eupyrene spermatozoa across the testicular basement membrane in the male

Each hypothesis should be examined, and the role of apyrene spermatozoa in sperm competition could be proposed.

Swallow and Wilkinson (2002) reviewed the putative role of selection in apyrene spermatozoa and grouped prior hypotheses concerning the possible function into three categories: (1) provisioning, (2) facilitation and (3) competition (Table 8.1), in addition to the replication error hypothesis that the apyrene spermatozoa is nonadaptive. Of three categories, provisioning is unlikely because apyrene spermatozoa appear to carry few nutrient reserves. Consequently, because there is a small nutritive role for apyrene spermatozoa to females, the major nuptial gift from the male must quantitatively be the spermatophore and accessory substances because of their mass.

The facilitation hypothesis includes the role of transportation. The apyrene spermatozoa might help eupyrene spermatozoa to transport from the site of production in the male testes to the site of the reproductive tract. Katsuno (1977) hypothesised that apyrene spermatozoa perforate the testicular basement membrane and thereby facilitate emigration of eupyrene sperm bundles from the testes to the ducts for ejaculation. In addition, because the mobility of eupyrene spermatozoa is lower than that of apyrene spermatozoa, the apyrene spermatozoa might assist the migration of eupyrene spermatozoa within the female reproductive tract.

Table 8.1 Hypotheses concerning the function of parasperm

Category	Hypothesis	Function
Nonadaptive	Nonadaptive	None
Facilitation	Transportation	Aid in emigration from testes
		Aid in migration within female
	Capacitation	Initiate capacitation
Provisioning	Provisioning	Provide nutrients to female
		Provide nutrients to eusperm
		Provide nutrients to ovum
Competition	Elimination	Remove/flush stored sperm
		Disable stored sperm
	Cheap filler	Delay re-mating
	Blocking	Block sperm entry
	Differential success	Both fertilisation competent
Other	Sex ratio	Alter sex ratio

Source: Swallow and Wilkinson (2002)

Before ejaculation and formation of a spermatophore, an eupyrene spermatozoa is immobile because of the bundles, although the apyrene sperm bundles have already dissociated. Then, the free apyrene spermatozoa might initiate promoting the dissociation of eupyrene sperm bundles by mechanically breaking the bundles apart, as well as mixing the highly viscous contents of the spermatophore (Swallow and Wilkinson 2002). In addition, the longevity of eupyrene spermatozoa might depend not only on the quality of each eupyrene spermatozoon but also on the environment inside the spermatheca, as pointed out by Konagaya et al. (2015). Maintaining sperm in the spermatheca during the lifespan has to be important for females of either monandrous or polyandrous species. In addition to the case of the polyandrous butterflies, therefore, high survival of apyrene spermatozoa found in the monandrous butterflies, such as the pipevine swallowtail, *Battus philenor*, appears to be a result of a suitable environment inside the spermatheca that enhances the survival of eupyrene spermatozoa.

Apyrene spermatozoa might be specialised to provide either a defensive or offensive advantage in competition between the ejaculates of different males. Four alternative hypotheses have been proposed under the assumption that sperm are in competition. The first is the *elimination hypothesis*. Because the apyrene spermatozoa are not involved in fertilisation but may serve to enhance the probability of fertilisation by eupyrene spermatozoa, apyrene spermatozoa seem to be preadapted for the seek-and-destroy mission of removing previously deposited eupyrene spermatozoa from competition, either by flushing a rival male's eupyrene spermatozoa from the spermatheca or by killing the eupyrene spermatozoa through direct interaction, probably in a 'kamikaze' role, sacrificing themselves in a 'war' with rival eupyrene spermatozoa for the good of their own ejaculate (Silberglied et al. 1984). Rival eupyrene spermatozoa must be effectively removed from competition, increasing the male's chance at fertilisation. In the armyworm, *Pseudaletia separata*, more apyrene spermatozoa may be an adaptation that serves to tackle the intensified competition from rival eupyrene spermatozoa (He and Miyata 1997).

Sperm displacement after repeat mating has been reported in several lepidopteran species (LaMunyon 2000). Etman and Hooper (1979) found, in the oriental leafworm moth, *Spodoptera litura*, that spermatozoa in the spermatheca of re-mated females started to decrease (to empty) just after the termination of copulation and then increased again. Pair et al. (1977) reported that a female of the hybrid between *Heliothis subflexa* and *H. virescens* decreased 88% of eupyrene spermatozoa in the spermatheca after re-mating.

Watanabe and Sasaki (2010) stated that spermatophore mass should be a good indicator of male quality, and the cryptic female choice consisting of spermatophore mass increases their fitness. In the swallowtail butterfly, *Papilio xuthus*, mated females, that is, just after re-mating, might decide whether to eject spermatozoa on the basis of the mass of the spermatophore from their last mate, if normal spermatophore mass would be a cue for cryptic mate choice by females. Because in the field the male can commonly produce 6 mg of spermatophore (Watanabe et al. 2000), females that received a spermatophore weighing less than 3 mg from the second mated male eject eupyrene spermatozoa before these arrive at the spermatheca. Consequently, the sons of a male with a low spermatophore production capacity may decrease the reproductive success of the female (Sasaki et al. 2015). Swallow and Wilkinson (2002) introduced one other potential and yet uninvestigated role of the apyrene spermatozoa, which trigger the elimination or flushing of rival sperm. The mechanism is that apyrene spermatozoa could act as a signal, possibly of a healthy eupyrene spermatozoa complement or as an indicator of superior genetic quality. Females of some species might evaluate before retaining or evacuating previously deposited spermatozoa.

Rutowski et al. (1981) stated that if a mated female re-mates before the migration of the first male's sperm to the spermatheca, the first spermatophore is pushed from the entrance to the rear of the bursa copulatrix, and then the second spermatophore is in the correct position to release sperm for sperm migration (*position effect*). The opening of the seminal duct provides only a restricted duct system to be available for the sperm to proceed from the spermatophore to the spermatheca. However, both types of sperm in the swallowtail butterfly, *Papilio xuthus*, migrate from the spermatophore to the spermatheca within 1 day of the termination of copulation (Watanabe and Hachisuka 2005). Therefore, because mated females commonly accept re-mating several days after the first mating, the spermatophore of the first male in the bursa copulatrix contains no sperm when re-mating. The spermatophore from the second male must be no barrier to sperm migration of the first male sperm from the spermatophore to the spermatheca.

Apyrene spermatozoa might function to block access to important areas of the female reproductive organs and exclude the successive male's eupyrene spermatozoa (*blocking hypothesis*). Watanabe et al. (2000) showed that apyrene spermatozoa of the swallowtail butterfly, *Papilio xuthus*, start to migrate from the spermatophore before eupyrene spermatozoa, probably because of its small size, resulting in a positional effect in the spermatheca for sperm competition but not in the spermatophore. Because the spermatheca has a single entrance with a cylindrical nature, the first male's eupyrene spermatozoa will be pushed away from that entrance by the second male's apyrene spermatozoa. Sperm from different males must be layered in the spermatheca, where apyrene spermatozoa form a barrier between preexisting sperm and the incoming eupyrene spermatozoa. Then, the second male's eupyrene spermatozoa will be in position to fertilise the eggs as they pass the entrance, and the apyrene spermatozoa block the eupyrene spermatozoa of the previously mated males for fertilisation when ovipositing. The adaptive significance of the early migration of a large number of apyrene spermatozoa from the spermatophore to the spermatheca may result in an increased confidence of paternity among males that deliver eupyrene spermatozoa, which migrate after the apyrene spermatozoa. Consequently, the last mate's eupyrene spermatozoa are the first to be used. However, positional effects in the spermatheca do not always occur, because a bimodal distribution of P_2 values with peaks near 0% and 100% is common in Lepidoptera (Simmons 2001). Sasaki et al. (2015) experimentally clarified that when the swallowtail butterfly females had mated twice, the male that transferred a larger spermatophore was more likely to be the principal sire, irrespective of whether he was the female's first or second mate. In addition, pushing sperm into the rear of the spermatheca would result in accumulation of spermatozoa in spermatheca with the number of matings, although LaMunyon (2000) reported that the number of the eupyrene spermatozoa in the spermatheca of re-mated females of the tobacco budworm, *Heliothis virescens*, is not significantly different from that of females mated once. Examination of the function of the spermathecal glands is needed.

Cook and Wedell (1999) stated that because of high mortality apyrene spermatozoa aid the transfer of eupyrene spermatozoa, in which case there should be a constant ratio between the two sperm types. The remaining latency of the female of the oriental armyworm, *Pseudaletia separata*, is correlated with the duration of apyrene sperm storage but not with storage of the longer-lived eupyrene spermatozoa (He et al. 1995). Apyrene spermatozoa that arrived in the spermatheca of the swallowtail butterfly, *Papilio xuthus*, also gradually disappeared within 2 days after copulation (Watanabe et al. 2000). Apyrene spermatozoa seem not to have remained in the spermatheca for long. When mating with mated females that have already been inseminated, therefore, apyrene spermatozoa must interfere with rival males' sperm in the spermatheca, filling the spermatheca and delaying female re-mating. The *cheap filler hypothesis* in which apyrene spermatozoa in the spermatheca reduce female receptivity for re-mating has been believed to be a dominant explanation of migration of apyrene spermatozoa to the spermatheca, although prerequisite on the hypothesis is that apyrene spermatozoa remain in the spermatheca for a long time after copulation.

Males of polyandrous species such as the swallowtail butterfly, *Papilio xuthus*, produced larger spermatophores but fewer eupyrene spermatozoa than males of the monandrous species, *Papilio machaon* (Watanabe and Kobayashi 2006), suggesting that there is a trade-off among ejaculate characters, because production of sperm rather than the spermatophore requires much nitrogen (Boggs and Gilbert 1979), and it is difficult to replenish nitrogen after eclosion in nectar-feeding butterflies (Marshall 1982). Although the rate of production of eupyrene sperm bundles of *P. xuthus* was lower than that of *P. machaon*, the rate of production of apyrene spermatozoa did not differ between the species. Assuming that there is a trade-off between ejaculate characters, this suggests that apyrene spermatozoa are important in sperm competition. Cook and Wedell (1999) found that re-mating females of the green-veined white, *Pieris napi*, have fewer apyrene spermatozoa already present in their spermatheca than females who did not re-mate, and stated that the transferring of a large number of apyrene spermatozoa functions as a 'filler' that delays female re-mating.

In the common grass yellow, *Eurema mandarina*, the number of apyrene spermatozoa in the spermatheca decreases about 2 days after copulation (Konagaya and Watanabe 2015). In contrast, the monandrous butterfly shows that a considerable number of apyrene spermatozoa still remain in the spermatheca after the collapse of the spermatophore, suggesting that apyrene spermatozoa could exist in the spermatheca throughout the female lifespan (Konagaya et al. 2015). Therefore, this supports the cheap filler hypothesis, although the proportion of apyrene spermatozoa to total sperm number decreased with age. He et al. (1995) pointed out that the "longevity" of apyrene spermatozoa is shorter than that of the eupyrene spermatozoa in the armyworm, *Pseudaletia separata*.

If apyrene spermatozoa function either as nutrient donations or in delaying the onset of female sexual receptivity for re-mating, the males may benefit by ejaculating large numbers of apyrene spermatozoa instead of a large spermatophore that is more costly for the experienced males.

Final Conflict of Apyrene Sperm

The spermathecal gland in females can have other functions, such as a storage and absorbing space for sperm, although the secretion of the gland serves as medium for the motile passage of spermatozoa (Drummond 1984). The apyrene spermatozoa migrate to the spermatheca earlier than eupyrene spermatozoa, filling in the spermatheca, and move out from the spermatheca to replace eupyrene spermatozoa because the spermatheca does not elongate as in the case of bursa copulatrix when the female mates more than once (Watanabe and Hachisuka 2005). Then, apyrene spermatozoa move to the spermathecal gland, are accumulated and then disappear, probably by absorption by ovipositing females. Thus, the destination of migrating apyrene spermatozoa is the spermathecal gland, although they are redundant sperm (Sivinski 1980). Cook and Wedell (1999) supported the cheap filler hypothesis in the green-veined white, *Pieris napi*, wherein females containing a larger number of apyrene spermatozoa in the spermatheca show lower receptivity for re-mating. For the armyworm, *Pseudaletia separata*, He et al. (1995) suggested that female receptivity for re-mating depends on the number of apyrene spermatozoa but not eupyrene spermatozoa in the spermatheca. Males of the Indian meal moth, *Plodia interpunctella*, transfer more apyrene spermatozoa when the risk of sperm competition is higher (Cook and Gage 1995). On the other hand, such a role for apyrene spermatozoa might be unnecessary in monandrous species because of the potentially low mating number. However, males of monandrous butterflies also produced apyrene spermatozoa. Vande Velde et al. (2011) showed that differences in sperm allocation strategies in males between polyandrous and monandrous species might be quantitative rather than qualitative, although the role of excess of apyrene spermatozoa arriving in the spermatheca has not been examined in monandrous species.

It can be understood that each spermatozoon has to be motile in the spermatheca after the sperm migration when alive. For mated females, assessing the viability of sperm in the spermatheca using the motility of sperm as an indicator must be important to fertilise eggs successfully. Oviposition could be stimulated by the presence of male ejaculates in the bursa copulatrix, as suggested by Watanabe (1988), and probably by the presence of mobile eupyrene spermatozoa in the spermatheca. Thibout (1979) reported that females show high receptivity for re-mating when motile eupyrene spermatozoa are absent in the spermatheca in the leek moth, *Acrolepiopsis assecella*. Cook and Gage (1995) suggested that motility of apyrene spermatozoa in the spermatheca was also important to reduce female receptivity for re-mating.

Activation of eupyrene and apyrene spermatozoa occurs in the spermatophore soon after copulation termination (Osanai et al. 1987), and both spermatozoa flow out of the spermatophore into the bursa copulatrix. Thereafter, both spermatozoa begin to migrate to the spermatheca, as first reported by Iriki (1941). Nabi and Harrison (1983) found that apyrene spermatozoa showed higher motility than eupyrene spermatozoa in the bursa copulatrix. Although both spermatozoa disappeared in the bursa copulatrix, migrating to the spermatheca, so small portions of both eupyrene and apyrene spermatozoa arrive in the spermatheca, as reported in the swallowtail butterfly, *Papilio xuthus* (Watanabe et al. 2000) and the cotton bollworm, *Helicoverpa armigera* (Yan et al. 2013). When arriving at the spermatheca, the apyrene spermatozoa as well as eupyrene spermatozoa are alive in the spermatheca about 1 week after the copulation, reported in several species such as the oriental armyworm, *Pseudaletia separata* (He et al. 1995), the green-veined white, *Pieris napi* (Cook and Wedell 1999) and the swallowtail butterfly (Watanabe et al. 2000). For the swallowtail butterfly, about 200 of 10,000 apyrene spermatozoa migrated remain in the spermatheca of the females 7 days after the copulation.

Although the presence of alive apyrene spermatozoa in the spermatheca about 1 week after copulation has been reported in several polyandrous species, high motility of apyrene spermatozoa soon after the migration to the spermatheca is found in the silkworm, *Bombyx mori* (Katsuno 1978) and the swallowtail butterfly, *Papilio xuthus* (Watanabe and Sasaki 2010). On the other hand, Nabi and Harrison (1983) observed not the moving but the gyrating of eupyrene spermatozoa in the spermatheca in the potato moth, *Phthorimaea operculella*. However, apyrene spermatozoa in the Chinese windmill, *Byasa alcinous*, lost their motility after arriving at the spermatheca, whereas the motility of eupyrene spermatozoa was maintained during at least 8 days (Konagaya and Watanabe, unpublished data). Because the cheap filler hypothesis requires high motility of apyrene spermatozoa in the spermatheca during the mating interval of females (Cook and Gage 1995), therefore the hypothesis could not be realised for the Chinese windmill, a monandrous species.

In polyandrous species, apyrene spermatozoa usually occupy more than 80% of sperm in the spermatophore (Swallow and Wilkinson 2002). For the swallowtail butterfly, *Papilio xuthus*, 94 % of sperm produced are apyrene spermatozoa (Watanabe et al. 2000). After ejaculation, a large number of apyrene spermatozoa compared to eupyrene spermatozoa is observed in the spermatophore. Although a part of both spermatozoa transferred arrive at the spermatheca, apyrene spermatozoa comprise the majority of sperm in the spermatheca. As a high percentage of apyrene sperm was observed in the spermatheca as well as in the male reproductive organs, the apyrene sperm migration therefore must have a specific role to increase the reproductive success of males.

One possible explanation for the function of apyrene spermatozoa is that they assist the migration of eupyrene spermatozoa from the bursa copulatrix to the spermatheca (Iriki 1941). Sahara and Takemura (2003) demonstrated that fertilisation in the silkworm, *Bombyx mori*, requires transferring of apyrene spermatozoa in the bursa copulatrix. Holt and North (1970) observed migration of eupyrene spermatozoa surrounded by apyrene spermatozoa in the reproductive ducts between the bursa copulatrix and the spermatheca. If apyrene spermatozoa have such a function, high motility and the sufficient number of apyrene spermatozoa in the reproductive ducts of females between the bursa copulatrix and the spermatheca after copulation may be predicted. On the other hand, Watanabe et al. (2000) reported that the earlier arriving of apyrene spermatozoa in the spermatheca than that of eupyrene spermatozoa in the polyandrous swallowtail butterfly, *Papilio xuthus*, would construct the layer of both types of sperm, resulting in the prevention of mixing of the eupyrene spermatozoa transferred from the first males. In addition, apyrene spermatozoa might carry nutrition to the spermatheca for females or eupyrene spermatozoa (Boggs 1995) because apyrene spermatozoa as well as eupyrene spermatozoa are protein rich in some species (Mancini and Dolder 2004).

The cheap filler hypothesis is not always the dominant explanation for the role of parasperm in insects. Swallow and Wilkinson (2002) pointed out that parasperm of diopsid flies cannot act as a cheap filler because of the low number of parasperm in one ejaculate compared to the capacity of the spermatheca. Although strong evidence for the cheap filler hypothesis is reported by Cook and Wedell (1999), Konagaya and Watanabe (2015) showed that apyrene spermatozoa disappeared in the spermatheca a couple of days after copulation in the common grass yellow *Eurema mandarina*, indicating a low possibility of the cheap filler hypothesis in some species. In addition, the cheap filler hypothesis on apyrene spermatozoa is not likely to apply to monandrous butterflies.

References

Boggs CL (1995) Male nuptial gifts: phenotypic consequences and evolutionary implications. In: Leather SR, Hardie J (eds) Insect reproduction. CRC Press, New York, pp 215–242

Boggs CL, Gilbert LE (1979) Male contribution to egg production in butterflies: evidence for transfer of nutrients at mating. Science 206:83–84

Cook PA, Gage MJG (1995) Effects of risks of sperm competition on the numbers of eupyrene and apyrene sperm ejaculated by moth *Plodia interpunctella* (Lepidoptera: Pyralidae). Behav Ecol Sociobiol 36:261–268

Cook PA, Wedell N (1999) Non-fertile sperm delay female remating. Nature (Lond) 397:486

Drummond BA III (1984) Multiple mating and sperm competition in the Lepidoptera. In: Smith RL (ed) Sperm competition and the evolution of animal mating systems. Academic, New York, pp 291–370

Etman AAM, Hooper GHS (1979) Sperm precedence of the last mating in *Spodoptera litura*. Ann Entomol Soc Am 72:119–120

Friedländer M (1997) Control of the eupyrene–apyrene sperm dimorphism in Lepidoptera. J Insect Physiol 43:1085–1092

He Y, Miyata T (1997) Variations in sperm number in relation to larval crowding and spermatophore size in the armyworm, *Pseudaletia separata*. Ecol Entomol 22:41–46

He Y, Tanaka T, Miyata T (1995) Eupyrene and apyrene sperm and their numerical fluctuations inside the female reproductive tract of the armyworm, *Pseudaletia separata*. J Insect Physiol 41:689–694

Holt GG, North DT (1970) Effects of gamma irradiation on the mechanisms of sperm transfer in *Trichoplusia ni*. J Insect Physiol 16:2211–2222

Iriki S (1941) On the function of apyrene spermatozoa in the silkworm. Zool Mag 53:123–124 (in Japanese)

Katsuno S (1977) Studies on eupyrene and apyrene spermatozoa in the silk worm, *Bombyx mori* L. (Lepidoptera: Bombycidae). II. The intratesticular behaviour of the spermatozoa after emergence. Appl Entomol Zool 12:236–240

Katsuno S (1978) Studies on eupyrene and apyrene spermatozoa in the silkworm, *Bombyx mori* L. (Lepidoptera: Bombycidae). VII. The motility of sperm bundles and spermatozoa in the reproductive organs of males and females. Appl Entomol Zool 13:91–96

Konagaya T, Watanabe M (2015) Adaptive significance of the mating of autumn-morph females with non-overwintering summer-morph males in the Japanese common grass yellow, *Eurema mandarina* (Lepidoptera: Pieridae). Appl Entomol Zool 50:41–47

Konagaya T, Mutoh N, Suzuki M, Rutowski RL, Watanabe M (2015) Estimates of female lifetime fecundity and changes in the number and types of sperm stored with age and time since mating in the monandrous swallowtail butterfly, *Battus philenor* (Lepidoptera: Papilionidae) in the Arizona desert. Appl Entomol Zool 50:311–316

LaMunyon CW (2000) Sperm storage by females of the polyandrous noctuid moth *Heliothis virescens*. Anim Behav 59:395–402

Mancini K, Dolder H (2004) Protein detection in spermatids and spermatozoa of the butterfly *Euptoieta hegesia* (Lepidoptera). Biocell 28:299–310

Marshall LD (1982) Male nutrient investment in the Lepidoptera: what nutrients should males invest? Am Nat 120:273–279

Nabi MN, Harrison RA (1983) Activity of sperm and fertility in the potato moth, *Phthorimaea operculella*. J Insect Physiol 29:431–435

Osanai M, Aigaki T, Kasuga H (1987) Energy metabolism in the spermatophore of the silkmoth, *Bombyx mori*, associated with accumulation of alanine derived from arginine. Insect Biochem 17:71–75

Pair SD, Laster ML, Martin DF (1977) Hybrid sterility of the tobacco budworm: effects of alternate sterile and normal matings on fecundity and fertility. Ann Entomol Soc Am 70:952–954

Rutowski RL, Long CE, Marshall LD, Vetter RS (1981) Courtship solicitation by *Colias* females (Lepidoptera: Pieridae). Am Midl Nat 105:334–340

Sahara K, Takemura Y (2003) Application of artificial insemination technique to eupyrene and/or apyrene sperm in *Bombyx mori*. J Exp Zool A 297:196–200

Sasaki N, Todoriki S, Watanabe M (2015) Effect of sperm ejection by females on male fertilization success in the swallowtail butterfly, *Papilio xuthus* L. (Lepidoptera: Papilionidae). J Res Lepid 48:39–44

Silberglied RE, Shepherd JG, Dickinson JL (1984) Eunuchs: the role of apyrene sperm in Lepidoptera? Am Nat 123:255–265

Simmons LW (2001) Sperm competition and its evolutionary consequences in the insects. Princeton University Press, Oxford

Sivinski J (1980) Sexual selection and insect sperm. Fla Entomol 63:98–111

Swallow JG, Wilkinson GS (2002) The long and short of sperm polymorphisms in insects. Biol Rev 77:153–182

Thibout E (1979) Stimulation of reproductive activity of females of *Acrolepiosis assectella* (Lepidoptera: Hyponomeutoidea) by the presence of eupyrene spermatozoa in the spermatheca. Entomol Exp Appl 26:279–290

Vande Velde L, Damiens D, Van Dyck H (2011) Spermatophore and sperm allocation in males of the monandrous butterfly *Pararge aegeria*: the female's perspective. Ethology 117:645–654

Watanabe M (1988) Multiple matings increase the fecundity of the yellow swallowtail butterfly, *Papilio xuthus* L., in summer generations. J Insect Behav 1:17–29

Watanabe M, Hachisuka A (2005) Dynamics of eupyrene and apyrene sperm storage in ovipositing females of the swallowtail butterfly *Papilio xuthus* (Lepidoptera: Papilionidae). Entomol Sci 8:65–71

Watanabe M, Kobayashi T (2006) Total sperm ejaculation in monandrous (*Papilio machaon*) and polyandrous (*P. xuthus*) swallowtail butterflies (Lepidoptera: Papilionidae) restricted to larval stage-derived nutrients. J Res Lepid 39:1–7

Watanabe M, Sasaki N (2010) Pattern of sperm storage and migration in the reproductive tract of the swallowtail butterfly *Papilio xuthus*: cryptic female choice after second mating. Physiol Entomol 35:328–333

Watanabe M, Bon'no M, Hachisuka A (2000) Eupyrene sperm migrates to spermatheca after apyrene sperm in the swallowtail butterfly, *Papilio xuthus* L. (Lepidoptera: Papilionidae). J Ethol 18:91–99

Wedell N, Wiklund C, Bergström J (2009) Coevolution of non-fertile sperm and female receptivity in a butterfly. Biol Lett 5:678–681

Yan S, Li H, Zhang J, Zhu J, Zhang Q, Liu X (2013) Sperm storage and sperm competition in the *Helicoverpa armigera* (Lepidoptera: Noctuidae). J Econ Entomol 106:708–715

Index

M. Watanabe, *Sperm Competition in Butterflies*, Ecological Research Monographs,
DOI 10.1007/978-4-431-55945-0

Index: Species of Lepidoptera

M. Watanabe, *Sperm Competition in Butterflies*, Ecological Research Monographs,
DOI 10.1007/978-4-431-55945-0

MIX
Papier aus verantwortungsvollen Quellen
Paper from responsible sources
FSC® C105338

If you have any concerns about our products,
you can contact us on
ProductSafety@springernature.com

In case Publisher is established outside the EU,
the EU authorized representative is:
Springer Nature Customer Service Center GmbH
Europaplatz 3, 69115 Heidelberg, Germany

Printed by Libri Plureos GmbH
in Hamburg, Germany